THE ATLANTIC SALMON
Natural History, Exploitation and Future Management

KU-460-147

THE ATLANTIC SALMON

Natural History, Exploitation and Future Management

W.M. Shearer

BSc, MSc, CBiol, FIBiol

A Buckland Foundation Book

Fishing News Books

Copyright © Trustees, The Buckland
Foundation 1992

Fishing News Books
A division of Blackwell Scientific
   Publications Ltd
Editorial offices:
Osney Mead, Oxford OX2 0EL
25 John Street, London WC1N 2BL
23 Ainslie Place, Edinburgh EH3 6AJ
3 Cambridge Center, Cambridge,
   MA 02142, USA
54 University Street, Carlton,
   Victoria 3053, Australia

All rights reserved. No part of this
publication may be reproduced, stored
in a retrieval system, or transmitted
in any form or by any means, electronic,
mechanical, photocopying, recording
or otherwise without the prior
permission of the publisher.

First published 1992

Set by Setrite Typesetters Ltd
Printed and bound in Great Britain by
the University Press, Cambridge

DISTRIBUTORS

   Marston Book Services Ltd
   PO Box 87
   Oxford OX2 0DT
   (**Orders**: Tel: 0865 240201
            Fax: 0865 721205
            Telex: 83355 MEDBOK G)

USA
   Blackwell Scientific Publications Inc
   3 Cambridge Center
   Cambridge, MA 02142
   (*Orders*: Tel: (800) 759 6102
            (617) 225 0401)

Canada
   Oxford University Press
   70 Wynford Drive
   Don Mills
   Ontario M3C 1J9
   (*Orders*: Tel: (416) 441−2941)

Australia
   Blackwell Scientific Publications
   (Australia) Pty Ltd
   54 University Street
   Carlton, Victoria 3053
   (*Orders*: Tel: (03) 347−0300)

British Library
Cataloguing in Publication Data

Shearer, W.M. (William MacDonald)
   The atlantic salmon: natural history,
   exploitation and future management.
   I. Title
   597.55

ISBN 0−85238−188−3

To the memory of my wife Lilias
who died before this book was completed

A PIONEER OF FISHERY RESEARCH

Frank Buckland 1826–1880

The man and his work

Frank Buckland was an immensely popular mid-Victorian writer and lecturer on natural history, a distinguished public servant and a pioneer in the study of the problems of the commercial fisheries. He was born in 1826, the first child of William Buckland DD FRS, the first Professor of Geology in Oxford who was an eminent biologist himself. From infancy Frank was encouraged to study the world about him and he was accustomed to meeting the famous scientists who visited his father. Like many other biologists of his day, he trained as a surgeon; in 1854 he was gazetted Assistant Surgeon to the Second Life Guards, having completed his training at St George's Hospital, London. He began to write popular articles on natural history and these were issued in book form in 1857 as 'Curiosities of Natural History'. It was an immediate success and was to be followed at intervals by three more volumes; although long out of print these can be found in second-hand bookshops and still provide entertainment and interest.

His success increased demands upon him as a writer and lecturer and he resigned his Commission in 1863. He had become interested in fish culture, then regarded simply as the rearing of fish from the egg. This involved the fertilization of eggs stripped by hand from ripe fish with milt similarly obtained. Release of fry was seen as a means of improving fisheries, particularly of salmon and trout, in rivers and lakes which had suffered from over-exploitation or pollution. He gave a successful lecture on the subject at the Royal Institution in 1863, subsequently published as 'Fish Hatching', and was struck by the intense interest aroused by his demonstration.

He was permitted to set up a small fish hatchery at the South Kensington Museum, the forerunner of the Science Museum, and by 1865 had collected there a range of exhibits which were to form the nucleus of his Museum of Economic Fish Culture. This aimed to inform the public about the fish and fisheries of the British Isles and for the rest of his life he laboured to develop this display. Although he was paid for his attendances at the Museum, the exhibits were provided by him at his own expense; in his will he gave the Collection to the nation.

National concern over many years at the decline of salmon fisheries, which suffered not only from overfishing and pollution but also extensive poaching and obstructions such as locks and weirs, led in 1861 to the passing of the Salmon Fisheries Act under which two Inspectors for England and Wales were appointed. When one of the original Inspectors resigned in 1867, Buckland was an obvious choice as successor. He had already accompanied the Inspectors on their visits to rivers and was also often asked for advice by riparian owners. He would think nothing of plunging into a river in winter to help net fish for the collection of eggs, a practice which probably led to his early death. Until 1866 he produced a steady stream of natural history articles mainly for 'The Field'. Then he helped to establish a rival journal, *Land and Water*, which he supported until his death.

Britain's growing population in the last century created many problems of food supply; the sea fisheries offered a cheap source of abundant first class protein and as a result the marine fisheries, and particularly the North Sea fisheries, grew spectacularly. Little was known about sea fish; no statistics of fish landings were available, at least in England, and the biological basis of fisheries was a mystery, though it was widely believed that marine fisheries were inexhaustible. Nevertheless there were disturbing indications that previously prolific fisheries were no longer profitable and many Royal Commissions were set up. The most famous was that of 1863, which had Thomas Henry Huxley as one of its members. Buckland himself sat on four Commissions between 1875 and his death, a fact which reflected his increasing standing as a fisheries expert.

During his lifetime a number of public fisheries exhibitions were held abroad, and he tirelessly pressed for something similar to be staged in the United Kingdom. Unfortunately he died before he could see his wish fulfilled, but there is no doubt that the exhibitions held in Norwich (1881), Edinburgh (1882) and London (1883) owed much to the public interest he had worked so hard to engender. It should be noted that the Marine Biological Association of the UK, with its famous laboratory at Plymouth, was a direct result of the enthusiasm and concern created by the Great International Fisheries Exhibition held in London in 1883.

He died in December 1880, possibly of a disease caught from his parrot, for he had always been careless about his health and must have worked for long periods at full stretch to maintain such a high output of material. What he wrote was sometimes uneven but he was often breaking new ground. At all times he was concerned to explain, to teach and, most particularly to make the general public aware of the importance of their fisheries and the need to protect and develop this great national asset. A few days before his death he signed his will. His wife was to have a life interest in his estate but he bequeathed a sum of money which on her death should be used to establish a trust fund to support 'a professorship of Economic Fish Culture, to be called The Buckland Professorship'. The main responsibility laid on The Buckland Professor was that lectures should be delivered each year at suitable venues in the United Kingdom.

The foundation

It is clear that Frank Buckland intended the term Fish Culture to be widely interpreted and to cover much more than fish hatching and the rearing of fry. Consequently, when the original £5000 endowment became available in 1926, after the death of his widow in 1921, the original Trustees of the Buckland Foundation took a broad view of the subjects that Buckland Professors should be invited to write and lecture about. In 1930, for example, the first Buckland lectures were given by Prof Garstang, a leading marine biologist of the time, on the subject of 'Frank Buckland's Life and Work'. The following year it was 'Salmon Hatching and Salmon Migrations' and after that 'The natural history of the herring in Scottish Waters'. As fisheries science and fishing methods evolved many subjects presented themselves which were unknown to Buckland and his contemporaries and succeeding Trustees have sought to ensure that such topics are covered so that the lectures have always been timely, important and of value and interest to those who depend for their livelihood on some aspect of fish and fishing. Each of the three Trustees holds office for a five-year period and the day-to-day business of the Foundation is managed by its Clerk.

In the Spring of each year there is a meeting of the Trustees at which the subject and the Buckland Professor for the following year is chosen. In accepting the invitation to hold office the nominee also accepts the responsibility for producing a text and giving at least three lectures at venues that provide the closest possible link with that area of fish and fishing being examined. The text has to be approved by the Trustees before the lectures are delivered. The Professor receives £700 for the delivery of the manuscript and £100, with expenses, for each lecture and a commemorative medal at the end of his year of office. More often than not the texts of the 43 Buckland Professors holding office so far have been published as books and copies of the more recent ones are available from Fishing News Books.

The Trustees feel that by continuing to keep alive, via the means willed to them through Frank Buckland's own inspiration, the memory of a man who dedicated his life to the improvement of the commercial fisheries of the British Isles they are helping, in their turn, to improve conditions in the present commercial fisheries. As the 50th Lecture in the series begins to come into view, they are hoping for increasing recognition of both Buckland, the Man and for the Foundation he instituted.

CONTENTS

PREFACE

In addition to being one of the most popular writers on natural history topics of the last century, Frank Buckland made significant contributions to the early development of both marine and freshwater fishery research and administration. His interest in and knowledge of salmon was recognized in 1867 when he was appointed Inspector of Salmon Fisheries for England and Wales. For the next 12 years, until shortly before his death, he worked tirelessly to arrest the decline in salmon stocks in many English rivers. The main cause of this decline was the increased demands on water by industry as it struggled to modernize. 'In England and Wales, as well as in Scotland,' Buckland wrote, 'manufacturers of all kinds of materials, from paper down to stockings, seem to think that rivers are convenient channels kindly given them by nature to carry away at little or no cost the refuse of their works.' One manufacturer claimed that sulphuric acid, which was released from his works, was a tonic for the fish.

In many ways, Frank Buckland's thinking was way ahead of that of his contemporaries and, indeed, of many of those who came after him. In a letter to *The Times*, summarizing the effect of a Salmon Fishery Amendment Act in 1873, he wrote 'After long experience, I am convinced that artificial culture of migrating Salmonidae can never compete with natural cultivation. This consists not in hatching out a few hundred thousand salmon eggs in boxes but in opening up for the parent fish as many miles of spawning ground as possible. The salmon know their own business much better than we do.' Later in the same letter, he continued 'Since the Salmon Act of 1861 was passed we have discovered that salmon will not obey the laws which the legislature enacted; and it has now been found necessary to call the salmon into consultation, and adapt legislation to the habits and instincts of these clever, mysterious but most valuable creatures. Salmon will not ascend rivers according to Act of Parliament, but will come up just when it pleases them to do.' I have no doubt, therefore, that Frank Buckland would have approved of my choice of subject for the series of lectures which formed the basis of this book.

Three lectures were given: at the Assembly Rooms, Wick, on 28th November 1989; at The Museum and Art Gallery, Perth, on 8th December, 1989; and finally at the Linnean Society, London, on 3rd April, 1990. On each occasion the subject was: 'The Atlantic Salmon; Natural History, Exploitation and Prediction.'

The practical management of salmon stocks to provide maximum catches without endangering the species' future and the development of fisheries are subjects which have interested me for many years. My particular interest has been in investigating the relationships between spawning stock, catch and recruitment. This book deals with these topics at a time when the level and manner of the exploitation and the management of the resource are changing rapidly. Management decisions are now taken at the international rather than the national level and the truly knowledgable amateur, whether netsman or angler, is having less and less say in the decision-making, even at the local level. An added complication has been the success of salmon farming which has had a major impact not only on the economics of the whole industry but also on the survival of native stocks in their present form. The conflicts which have arisen between anglers and netsmen over the claim by some anglers that nets have been removing too high a proportion of the larger fish have not benefited the salmon; they have tended to mask the real problems, such as degradation of habitat and the impact of fish farming, and have given the administrators an excuse to take no action.

Research into the salmon populaton of the R. North Esk has now reached a stage at which sufficient data are available to allow the fate of adult salmon derived from particular spawnings to be quantified and the fisheries which exploit these salmon to be identified. Thus, the possibility of modelling the R. North Esk stock dynamics is no longer a pipe dream and the advantages which can be derived from this new approach when fixing catch quotas and taking management decisions are discussed in some detail.

Evidence is presented which indicates that smolt production is not limited by extensive exploitation by fisheries on the high seas or in home waters but by the loss and degradation of the nursery habitat which frequently follows changes in land-use. The book ranges widely over all aspects of this most important topic.

Stages of the life history where additional research is required have been identified and the magnitude of the annual budget thought necessary to investigate the present increased natural smolt mortality in the sea is given merely as an example of the funds necessary to extend knowledge to the marine environment. However, it is suggested that this sum should not be viewed in isolation but in the context of the estimated capital value (£17 000) of a single rod-caught salmon in Scotland.

Much of the knowledge incorporated in this book has been acquired over almost half a century of contacts with owners and lessees of net fisheries, netsmen and friends and colleagues in the salmon fishing industry and in salmon research both in Scotland and across the North Atlantic. To them all, I owe a tremendous debt of gratitude. I am particularly grateful to a number of them for help in preparing some of the data and the provision of illustrations. Mr Bryce Whyte aged most of the fish. All the line diagrams, with the exception of those describing the R. Thurso data which were produced by Mr Bill Hall, were produced by Mrs J.M.A. Milne. Mrs M. Gammie and Mr A. Rice provided valuable assistance with the artwork. Mr Tom McInnes was responsible for most of the photography. The use of other illustrative material is acknowledged in the appropriate places and I am obliged to Messrs Croom Helm and Messrs Chapman and Hall, the publishers of 'Atlantic Salmon:

Planning for the Future' and 'Ecology and Management of Atlantic Salmon' respectively for their permissions to reproduce this illustrative material. I am also grateful to the General Secretary of the International Council for the Exploration of the Sea for permission to quote from their publications.

Early versions of the manuscript were read by Mr D.A. Dunkley, Freshwater Fisheries Laboratory, Field Station, Montrose, a respected colleague for more than 20 years, Mr A.V. Holden, one-time Director of the Freshwater Fisheries Laboratory, Pitlochry, Professor David Jenkins, one-time Director of the Institute of Terrestial Ecology Laboratory, Banchory, and Professor A.D. Hawkins, Director of Fisheries Research for Scotland, Aberdeen, and I am deeply indebted to them for their wise comments and helpful suggestions. In addition, I called on Mr Dunkley for help and advice on many occasions during the preparation of the text for publication and I always found his comments extremely useful.

I am also indebted to Mr Bill Malcolm, Scottish Office Agriculture and Fisheries Department; Mr Nick Brown, Ministry of Agriculture, Fisheries and Food, Mr Ruck Keene, Crown Estate Office and Dr Carlos Garcia de Leaniz for so kindly taking time to read some of the sections and making a number of most valuable comments and suggestions.

Data describing catches taken in the Rivers Thurso, Spey, Dee, Tweed, Tyne, Avon, Stour, Exe, Ribble and Lune, in some instances covering a period of more than one hundred years, were provided by Viscount Thurso of Ulbster; Mr Robert Clerk, Messrs Smiths Gore; Mr James Scott, Aberdeen Harbour Board; Mr James Reed, Berwick Salmon Fisheries Company; Mr Tony Champion, National Rivers Authority, Northumbrian Region; Dr D.R. Wilkinson, National Rivers Authority, Wessex Region and Mr Ted Potter, Ministry of Agriculture, Fisheries and Food, Fisheries Laboratory, Lowestoft and I acknowledge their help with gratitude.

It gives me great pleasure to thank the Esk District Salmon Fishery Board, the proprietors and lessees of river and coastal fishings along the R. North Esk and on the adjacent coast and Messrs Jos. Johnston and Sons Ltd., Montrose; much of the research centred on the R. North Esk would not have been possible without their permissions, help, understanding and encouragement. Thanks in no small measure are also due to the many netsmen around the Scottish coast who willingly gave permission to tag and release salmon caught in their nets and subsequently returned details of any tagged fish which they caught.

Finally, I am most grateful to the Trustees of the Buckland Foundation for inviting me to be the Buckland Professor for 1989. It was a great honour to follow in the footsteps of the late Mr W.L. Calderwood and the late Mr W.J.M. Menzies, the late Mr F.T.K. Pentelow and the late Mr Arthur E.J. Went, all highly respected Inspectors of Salmon Fisheries in Scotland, England and Wales and Ireland, whose wise advice during their respective appointments left viable fisheries, both net and rod, and salmon stocks at a satisfactory level.

William M. Shearer
Montrose,
Scotland.

CHAPTER 1

INTRODUCTION

It has been said that anyone who has not seen a wild salmon has not seen what a fish should be − a statement which tells us more about the species than any detailed description. The wild Atlantic salmon's shape is hydro-dynamically perfect, streamlined and ideally proportioned, which allows it to inhabit both river and sea with ease. The cryptic coloration of the juvenile salmon in fresh water provides perfect camouflage, as does the spawning livery of the adult after its return from the sea. In the sea, the salmon's deep aquamarine back, silver sides and white underside all add to the attraction of the species, this classic counter-shading camouflage (typical of a pelagic species) being complemented by its speed, agility and endurance (Plate 1 in the colour section).

On two counts, the species was thought in ancient times to have mystical qualities. These are, first, its apparent ability to appear and disappear at will, and secondly, its ability to surmount substantial obstacles. In northern European rivers, at one moment thousands could be seen relentlessly swimming upstream and the next they were gone. At falls, they could be seen repeatedly jumping as high as 3 m or more (10−12 ft). This latter ability earned them the name 'the leaping fish' a title eventually recognized in their scientific name *Salmo salar* − salmon the leaper.

It was thought that no fish could possibly leap so high without some supernatural assistance or power. The scientific sages of the day, however, dispelled this belief and explained the dynamics of the leap by saying that the salmon took its tail in its mouth and by rotating its body like a wheel was able to convolute over the most difficult waterfall (Dunfield 1985). Today, we have considerably more knowledge of how salmon leap because their habits and life history have been studied intensively for over a century.

This book describes the life history of the Atlantic salmon, highlighting the published results of the last 30 years of research. It details all aspects of the fisheries which exploit this most valuable resource and it discusses the present status of the Atlantic salmon in the UK with special reference to Scotland. Finally, it offers some views on the future for salmon in the UK and in Scotland in particular.

It is believed by some scientists that the ancestor of salmon evolved with the group of higher bony fishes which appeared in the Cretaceous Period, 70 million years ago (Ommanney 1963). At this time, continental drift would not yet have separated the land masses of Europe, Greenland and North America.

The subsequent distribution of the species would have been facilitated both by a tendency for the growing population to enlarge and extend its range, and by the drifting apart of the land masses. If the species was basically marine, its principal feeding areas in the sea may have remained relatively stationary. But as the continents continued to drift apart, the migration routes to and from these areas would have become more extended as the species continued to expand its range (Dunfield 1985). This theory may explain the present use of a common marine feeding ground at west Greenland by salmon from both sides of the Atlantic whose natal streams are now widely separated. Others believe, however, that the species was basically marine and also developed its anadromous habits during the same period of glaciation.

An alternative view, however, is that the salmon was originally a fresh water fish which was forced into the marine environment by the massive terrestrial ice sheets which advanced over its range.

Although in the 19th century salmon could be found in every country whose rivers drained into the North Atlantic Ocean and the Baltic Sea, they have subsequently disappeared from such major rivers as the Rhine. The present distribution of the Atlantic salmon stretches from Massachusetts, USA, in the west through the eastern seaboard of Canada across to Greenland (where salmon spawn in one river, the Kapisillit), to Iceland, the Faroes, Norway, Sweden, Finland, eastward to the Pechora river in Russia, westward again to the Baltic Sea and the rivers of France which flow into the Bay of Biscay and the English Channel, and then to the British Isles, including the United Kingdom and the Republic of Ireland. The species reaches its southern limit in Spain and Portugal.

The main advances in knowledge of salmon in the last 30 years have been not only concerning their movements in the sea but also their population structure at the river stage. The main event has been the success of salmon farming and its major impact on the economics of the whole industry. Research into the biology of the Atlantic salmon has been based on studies centring on spawning behaviour, smolts and smolting, trapping smolts and adults just above the head of an estuary, and investigating the feeding grounds of salmon at sea. Exploitation of salmon has been controlled by international regulation of marine fisheries of salmon and agreement of total allowable catches (TAC) at sea fisheries in some areas.

Investigations into the biology of the salmon at the spawning stage have revealed the importance of the amount and quality of the nursery area in each river and the distribution of redds. These factors are more likely to determine the annual production of young salmon than the number of spawning fish. Much knowledge has been accumulated on the life history of different cohorts of smolts. These usually stay in the river for up to four, or very occasionally five, years in Scotland, although this stage in the life history may last much longer in the higher latitudes, such as northern Norway and Labrador where parr may remain in fresh water for eight or more years before migrating to sea. Some fish may become smolts as early as during their first year, while some parr become precociously mature and capable of fertilizing salmon eggs without going to sea at all. Smolts which stay in the river longest tend to return to the river in the first half of the year. Similarly, salmon which have spent longest in the sea also tend to return to their natal river early in the year.

The lengths of time for which smolts stay in the river, and salmon stay in the sea, are probably genetically determined, but these time periods are also influenced by rate of growth and the availability of food in the fresh water and marine phases of their lives. However, details of the underlying physiological mechanisms are largely unknown and this explains one of the big gaps in the salmon story.

Much research has been devoted to analysing trends in salmon catches by netsmen and anglers. Such analyses are often difficult because of the lack of useful data on fishing effort; moreover, even where such data are available, interpretation of the results has been greatly complicated by differences in the sea age at which fish return and in the times of year of return. While salmon may return in any month of the year, they can be divided broadly into those which return during the winter and those which do so during the summer. Fish returning in the winter, known as spring fish, have usually spent more time in the sea than fish which return — usually after only one winter in the sea — in the summer. The latter class of fish tend to have increased in number while multi sea-winter (MSW) fish have greatly declined. Some of the larger rivers may have a third group of fish which 'run' in the late summer and autumn, and these are also MSW fish. In many rivers, the increase in the number of grilse (1 sea-winter (SW) fish) has compensated for the decrease in the number of MSW salmon so that there may not have been a great change in the total number of salmon overall.

Conflicts have arisen between anglers and netsmen over the claim by anglers that nets have been removing too high a proportion of the scarce MSW fish which return in late winter. This has led to the continuing removal of nets from rivers by anglers in the belief that this will increase the availability of the large MSW fish to the rod fishery.

However, the length of time spent in the sea — a period which is related to the time of return but complicated by the length of time spent by fish in the river as smolts — is also associated with the temperature of the sea in the areas where salmon feed and grow. This stage of the life history of the salmon is inadequately understood and is a major area for future research.

Little is known about how the fish navigate at sea or the routes by which they return to their natal rivers. However, much work has been done on the movements of fish in British coastal waters and this work will be described in detail. Salmon tend to return to the rivers which they left as smolts and the time of entry to the river by adults is dependent upon where in the system the fish will eventually spawn. The earlier the fish enter the river the further upstream they tend to move; fish which enter in the winter tend to spawn in the upper reaches while those which enter in the autumn spawn low down in the main river. As a result, the sea age-groups tend to be isolated at spawning time: MSW fish mainly spawn in the upper reaches and the tributaries entering the main river in this area and grilse mainly spawn in the middle and lower reaches and in the associated tributaries.

It follows that if MSW salmon decline in numbers, as has happened in recent years, anglers who own the upper beats of a river will experience the greatest decline in fish availability because their fishing depends largely on this group of fish. It is not known whether MSW fish are a genetic strain of the salmon, tending to breed other MSW fish when two of them mate together, and whether different spawning areas are used by different genetic components of the salmon stock. This is another big area for future research. We need to know more about the interaction between inherited characteristics and the influence of environmental factors including food in the river and food in the sea.

There is no evidence that the Faroes and Greenland fisheries are responsible for the big decline in the numbers of MSW fish entering British rivers; these two sea fisheries had some impact on salmon stocks but this occurred after the decline of MSW fish had begun. Nor has the number of MSW fish entering rivers shown a significant increase following the greatly reduced marine fisheries catch as a result of quotas and, more recently, because of the economic pressure on the industry arising from the large number of farmed fish released on to the market.

It is unlikely that the cessation of net fishing in many Scottish rivers will result in an increase in the number of MSW fish available for anglers in the near future. The loss of this class of fish is a biological event and is not caused by overfishing. If environmental conditions revert, more spring fish will be expected. In the meantime, however, greater understanding is needed of the importance of the different genetic characteristics which partly determine where fish will spawn within a river and how long they may spend in the river and in the sea. Fishery owners need to improve and maintain the improvement of habitat, particularly in the upper stretches of rivers, by preventing siltation and bank erosion. Taking such action is the most important management activity for proprietors now and in the foreseeable future.

CHAPTER 2
THE LIFE HISTORY

The main features of the life history of the Atlantic salmon have probably been known for at least 450 years. They spawn in fresh water where the young fish spend some time before migrating to the sea. Once in salt water, they grow more rapidly and migrate long distances. It is somewhat surprising that the stages now known as parr and grilse were not positively identified until the middle of the 19th century. Even today, some older fishermen on the R. Tweed, for example, are adamant that grilse are a separate species. However, their views may not be as wide of the mark as initially thought because most salmon biologists now concede that each salmon river has its unique stock of salmon and the larger systems may have several stocks, of which grilse may be one.

Allan & Ritter (1975) described the various life-history stages on both sides of the Atlantic (Table 2.1). Seven stages are defined: alevin, fry, parr, smolt, post-smolt, salmon and kelt. The first three stages are found only in fresh water, while the other four stages are common to both salt and fresh water. Salmon are termed anadromous because they spawn in fresh water and migrate to the sea to feed. Eels (*Anguilla anguilla*), on the other hand, are catadromous because, unlike salmon, they spawn in the sea and migrate to fresh water to feed.

In Scotland, and to a lesser extent in other parts of the UK (mainly as a result of pollution), most rivers with free access to the sea and with no insurmountable obstruction support populations of Atlantic salmon. Salmon migrate from the sea into many of these rivers in every month of the year. In some of the larger rivers, some fish entering in October in one year may stay in fresh water until the October of the following year before spawning. Salmon usually spawn in gravelly areas between mid-October and late February. The female selects the spawning site and the location is normally where salmon spawned in previous years (Webb & Hawkins 1989). Some females may spawn at more than one site, and displacement (overcutting) of existing completed redds may occur.

Juvenile salmon usually spend one to four years, occasionally five in Scotland, in fresh water developing through alevin, fry and parr stages, and finally becoming smolts (Plates 2 and 3 in the colour section). This is the stage at which salmon go to sea. Smolts migrate down the rivers into the sea and subsequently travel considerable distances (e.g. to the Norwegian Sea or Greenland areas). In Scotland, the spawning season is generally earlier than in

Table 2.1 Terminology presently in use for Atlantic salmon (*Salmo salar* L.).

Term	Definition
Alevin	Stage from hatching to end of dependence on yolk sac as primary source of nutrition.
Fry	Stage from independence on yolk sac as primary source of nutrition to dispersal from redd.
Parr	Stage from dispersal from redd to migration as a smolt: 0+ parr, parr less than 1 year old (parr of the last year's hatch); 1+ parr, parr 1 year or over but less than 2 years; 2+ parr, parr 2 years or over but less than 3 years; 3+ parr, parr 3 years or over but less than 4 years; precocious parr, male parr fully ripened or matured in freshwater; partially silvered-parr, parr that are partially silvered and migrating downstream prior to the normal smolt run.
Smolt	Fully-silvered juvenile salmon migrating to the sea.
Post-smolt	Stage from departure from the river until onset of wide annulus formation at the end of the first winter in the sea: 'pre-grilse', post-smolt stage returning to freshwater to spawn in year of smolt migration.
Salmon	All fish after onset of wide annulus formation at the end of the first winter in the sea: (a) 1-sea winter (1SW) salmon, salmon which has spent one winter at sea (grilse when maturing to spawn); (b) 2SW salmon, salmon which has spent two winters at sea; (c) 3SW salmon, salmon which has spent three winters at sea; (d) 4SW salmon, salmon which has spent four winters at sea; (e) previous spawner, salmon which has spawned on previous occasion(s).
Kelt	Spent or spawned-out salmon until it enters salt water.

Source: Allan & Ritter (1975)

the southern parts of the UK while the average period which the juveniles spend in fresh water prior to smolting tends to be longer. One-year-old smolts are relatively uncommon in most Scottish rivers, while in rivers draining into the English Channel they form the major age group. The mean length of Scottish smolts at migration is about 12.5–14.0 cm (range 9.0–22.0 cm). Because of the growth added by the younger smolts in the months immediately before they migrate, particularly during the second half of May and early June after the older smolts have migrated to sea, the variation in length between smolts migrating at different ages is small. However, Scottish smolts are generally shorter at the same age than those produced in English rivers, particularly those situated in the southern half of England. Smolts grow rapidly in salt water, some increasing their weight from approximately 30 g to 5 kg in the 15 months before they return after one or more winters at sea to the rivers to spawn. In Scotland, less than 5% normally survive to spawn a second time (Mills 1986) (Fig. 2.1).

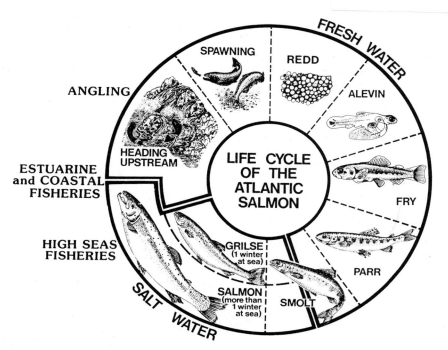

Fig. 2.1 Life cycle of Atlantic salmon.

CHAPTER 3

THE R. NORTH ESK AND
ITS RESEARCH FACILITIES

Much of this book is based on the results of research in the R. North Esk
where there is a range of research facilities including a trap for smolts and
adult salmon and an automatic fish-counter (Fig. 3.1). In addition, the owners
of the net fishery give free access to their catches and to their catch data and
the District Salmon Fishery Board together with the owners of the various rod
and net fisheries on the adjacent coast and along the length of the river are
sympathetic to research and have been most helpful and co-operative.

3.1 The R. North Esk

The R. North Esk, in north-east Scotland, is one of Scotland's major salmon
rivers. It enters the North Sea almost midway between Aberdeen and Dundee
(Fig. 3.2). It is a small river system, having a drainage area of about 732 km^2
and a total length of about 50 km. Discharge in the R. North Esk ranges
between 2 and 280 m^{3}s^{-1}, with an average daily flow of 13 m^{3}s^{-1}. Its topmost

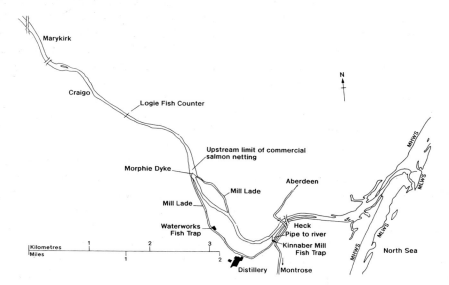

Fig. 3.1 Location of various facilities on the R. North Esk associated with the research
on Atlantic salmon in that river.

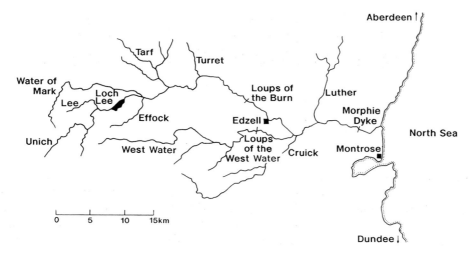

Fig. 3.2 The R. North Esk.

tributaries rise in Invermark Forest at a height of about 750 m. Two of the main tributaries, the Lee and the Unich burns, unite above Loch Lee. Approximately 1.5 km below the loch, another large tributary, the Water of Mark, joins the outflow to form the R. North Esk. Other major tributaries numbered among the 75 tributaries in the catchment area include the West Water, Cruick and Luther. The large area of good spawning gravel which these streams provide probably accounts for the position of the R. North Esk as the major east-coast salmon producing river measured on the basis of catch per kilometre.

Two major natural obstructions to the free passage of migratory fish are located above the village of Edzell. These are the Loups of the Burn on the main river and the Loups of the West Water (Fig. 3.2). The West Water falls were first blasted and substantially improved in 1947 and the fish pass, cut through the solid rock adjacent to the Loups of the Burn, was completed in 1949 (Plate 4). Minor improvements, together with the easing of less difficult natural obstacles to the free passage of salmonids, have since been made at both sites, the latest in the early 1980s. In the past, there were several weirs across the river but only one remains, Morphie Dyke, some 4.5 km from the sea and *c* 2 km above the head of tide at high water spring tide. This 'V'-shaped dyke has a fish pass, rebuilt in 1950, incorporated at its apex. The approach to the fish pass was improved in 1981 (Plate 5 in the colour section).

Water for domestic and industrial supplies is abstracted at two points in the river. In the headwaters, Loch Lee has been raised by a low barrier to form a reservoir. In the lower reaches, Kinnaber Waterworks on Kinnaber Lade draws water from the river at Morphie Dyke to provide the water supply for the town of Montrose. This latter abstraction ceased in 1990. During summer months, water for irrigation is pumped from the river at a number of sites, the volume and number of sites depending upon the degree of drought and the crops being grown.

Plate 4 Fish pass at the Loups of the Burn.

There has been little forestry development of the catchment area in the R. North Esk. Small areas have been planted in moorland in the headwater glens, principally as deer (*Cervus elaphus scoticus*) shelters rather than for timber. Scattered, largely relict woods of birch (*Betula pendula*) and alder (*Alnus glutinosa*) line the river banks below the confluence of the Water of Mark and the Lee as far downstream as the Loups of the Burn where beech (*Fagus sylvatica*) predominates. Having now crossed the Highland Boundary Fault Line, the river flows to the sea through the rich agricultural land of the Howe of the Mearns.

The water quality of the R. North Esk is good (Table 3.1), principally because there is no major industry, apart from two distilleries, sited within the watershed. However, because the lower reaches of the river flow through rich farming land, there is an occasional discharge of silage effluent. Infrequent fish kills can be associated with the inadvertent release of chemicals used in agriculture. In addition, the demand for more arable land has led to major drainage schemes which, apart from altering flow patterns, have destroyed nurseries of juvenile salmon. This is because the ditches and smaller feeder streams have been straightened and deepened to become more efficient water carriers. At the same time, soil erosion has been accelerated both as a result of the improved drainage and the increased proportion of the year when the top soil is bare and exposed to wind. Much of this eroded soil ultimately lands in the water course where it eventually forms a mat of silt over the gravel.

Table 3.1 Quality of R. North Esk water in 1985 and 1986.

Year	Temp (°C)	pH value	Dissolved oxygen (%)	Biochemical oxygen demand (mg l⁻¹)	Suspended solids (mg l⁻¹)	Ammoniacal nitrogen (mg l⁻¹)	Total oxidized nitrogen (mg l⁻¹)	Alkalinity (mg l⁻¹)	Chloride (mg l⁻¹)	Conductivity (μs cm⁻¹)	O-phosphate (mg l⁻¹)	Water quality index value
1985 Max	13.6	8.0	105.0	4.8	8	0.11	3.78	37.5	13.5	169	0.20	93
Mean	7.8	7.4	98.2	2.5	3	0.03	2.18	28.7	10.7	140	0.04	86
Min	2.5	7.1	93.6	0.0	0	0.00	1.36	13.5	7.0	100	0.01	66
1986 Max	16.4	8.8	123.0	3.2	11	0.10	3.43	45.0	16.0	218	0.07	93
Mean	7.6	7.6	101.1	1.8	5	0.03	1.75	27.7	11.0	135	0.03	89
Min	1.9	6.9	87.2	0.3	1	0.00	0.75	10.0	7.0	89	0.01	83

Source: Tay River Purification Board's Annual Reports for 1985 and 1986

3.2 The fishery

The Atlantic salmon population in this river has been extensively studied since 1964, centring on the parr (those >9 cm in length in two selected upper tributaries), the smolts just before they leave the river, and the adults on their return journey from the sea.

The North Esk Salmon Fishery District (Fig. 3.2) is among the most heavily fished areas for salmon in Scotland although in recent years, netting effort, both in the river and on the coast within the fishery district, has decreased. Outside the estuary, salmonids are caught using fixed engines (bag nets and fly nets) but the only permissible methods within estuary limits are net and coble and rod and line.

3.2.1 Net and coble

Although there are 22 recognized netting sites in the lower 7 km or so of the R. North Esk, net and coble fishing is now restricted to the lowermost 5 km (Plate 6). In this region, netting is concentrated at two sites, Morphie Dyke and the Nab, the other stations being fished only periodically (Fig. 3.3). Although netting is permissible from 0600 h on Monday until 1800 h on

Plate 6 Net and coble fishing (sweep-netting).

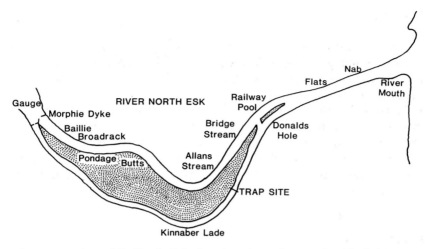

Fig. 3.3 Lower reaches of R. North Esk showing sites of net and coble fishing stations.

Friday (a 108-hour period), fishing is restricted to about 20% of this time. In recent years, the netting effort has been markedly reduced, the majority of the sites formerly netted now being let to anglers.

3.2.2 Fixed engine

Netting shifted from the rivers to the coast around 1830. This led to the design of a completely new generation of nets — the fixed engines. The term includes bag nets, fly nets and jumper nets (Plates 7, 8 & 9). A maximum of 35 fixed engines is now operated in the North Esk Salmon Fishery District at three stations. This represents a decrease of about 60% in fixed-engine fishing effort in the District since 1972. Netting does not now begin until mid-March and the maximum fleet of nets operates only from mid-June to mid-August. Estuary limits, within which no fixed engine can be set, are defined annually at the equinoctial spring tide.

3.2.3 Rod and line

Rod fishing for salmon is carried out along the whole length of the river from the lower reaches right up into the headwaters. Most of the fisheries are let (Plate 10).

Plate 7 (a) Fishing a single-headed bag net.

Plate 7 (b) A double and a single headed bag net.

3.3 Smolt and adult traps

The original smolt trap was installed at Kinnaber Waterworks (Fig. 3.1). A set of vertical screens was installed in the lade to divert fish into one of the coarse gravel filter beds of the waterworks (Shearer 1972). The use of this trap was discontinued in 1969, and in 1970 a custom built trap (Fig. 3.4) was constructed about 1 km downstream at the site of the former Kinnaber Mill, still on Kinnaber lade (Fig. 3.1). The amount of water drawn into the lade is regulated by a sluice to approximately 1.6 million litres per day. During drought conditions, this amount is greater than that flowing in the 2 km stretch of the

Plate 8 A fly-net.

main river between the intake and the outlet of the lade, but it is only a small proportion of the total flow when the river is in spate. Kinnaber Mill trap is of simple design, consisting of a fence across the lade and a holding tank into which downstream migrants diverted by the fence are collected and held until

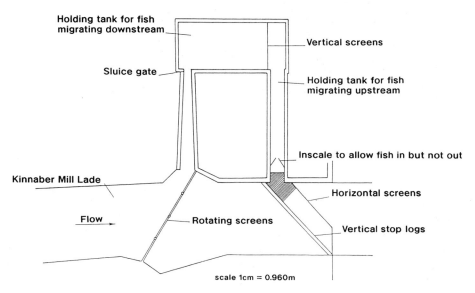

Fig. 3.4 Diagram of Kinnaber Mill fish trap.

Plate 9 A jumper net.

release. The fence is made up of four self-cleaning screens, which automatically rotate when a pre-set water level is exceeded. The gap between the slats of the screen is 13 mm. Upstream migrating fish are diverted by stop-logs through an inscale into a separate holding tank. Fish are prevented from surmounting the barrier of stop-logs by fixed horizontal screens (Plate 11 in the colour section). Upstream and downstream migrants in the holding tanks are separated by screens with 10 mm gaps between individual slats. The construction of the tank allows it to be rapidly drained, and captured smolts can be diverted into small holding boxes of water which can be lifted from the trap by a powered winch. Adults are individually caught in a custom-made, plasticized canvas bag (Fig. 3.5). This system reduces fish handling to a minimum. Prior to 1970, adult fish were caught for research purposes in Kinnaber Lade by electro-fishing.

The main limitation of this sampling system is that the number of fish taken in the trap is not directly related to the density of fish moving in the river. This is because the proportion of the total flow drawn into the lade varies with the water level at Morphie Dyke (Plate 5 in the colour section).

3.4 The automatic fish counter

This tool has, for the first time, permitted the number of potential spawners ascending a river upstream of a fixed point to be counted accurately.

Plate 10 Angling.

Since 1981, the fish which escaped capture in the lower R. North Esk have been automatically counted as they cross a specially built, three-channel, two-stage, compound Crump weir at Logie some 6 km from the sea (Plate 12 in the colour section). The weir is of novel construction, comprising trench sheeting foundations embedded in the gravel substrate of the river and pre-formed glass-reinforced plastic (GRP) deck sections supported on triangular steel bearers (Brown 1981).

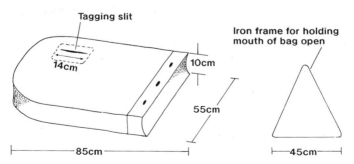

Fig. 3.5 Canvas bag for constraining and tagging adult fish caught in Kinnaber Mill trap.

Until 1986, three resistivity fish counters produced by the North of Scotland Hydro-Electric Board (NSHEB) were used, one connected to each section of the three-channel weir. Despite problems with spurious counts resulting from the action of ice, or wind-driven waves crossing the weir, good results were obtained as a consequence of twice-daily calibration of the counters. The performance of the counters was regularly checked using an independent closed circuit television (CCTV) surveillance system. An analogue chart recorder was used to record the signal waveforms detected (Fig. 3.6) when fish crossed the electrode array attached to the downstream face of the weir and connected to the counter (Plate 13).

Since 1986, the 'Logie' counter has been used. This instrument was developed jointly between the Department of Agriculture and Fisheries for Scotland and Aquantic Ltd, a relatively small company in Dingwall, Ross-shire, who include

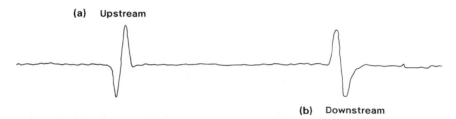

Fig. 3.6 Waveform produced by a salmon migrating upstream (a) and downstream (b) over the electrode array attached to the Logie fish counter.

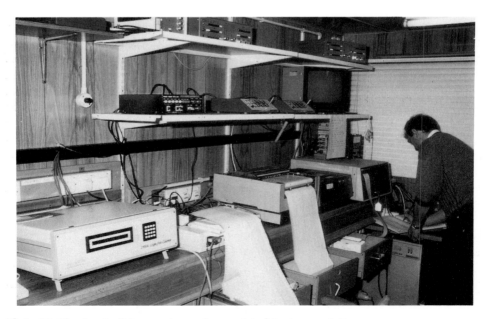

Plate 13 The Logie fish counter and associated instrumentation.

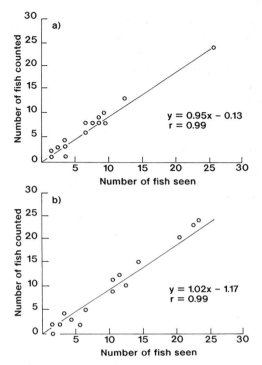

Fig. 3.7 Relationship between the numbers of fish seen on closed circuit television and the fish counted in the 1987 trials of the Logie fish counter: (a) refers to upstream migrants; (b) refers to downstream moving fish.

the designing and construction of electronic equipment among their activities. Basically, all counters of the 'resistivity' type detect fish by the transient reduction they produce in the resistance measured between adjacent pairs of stainless steel electrodes installed across a river. Using a powerful algorithm, the 'Logie' counter compares signals detected with a reference signal to determine whether they conform to fish signals (Fig. 3.6). The counter calibrates itself every 30 min. to accommodate changes in water conductivity, cable resistance, electrode resistance and water temperature. Because three electrodes span the river, the direction of movement of each fish crossing the weir can be determined. Tests using CCTV apparatus have shown that the reliability of the counter exceeds 95% for upstream and downstream migrants, both of which are counted and logged separately (Fig. 3.7).

The one real objection to this method of counting fish is that it can delay the upstream migration and the fish may congregate below the structure. Exhaustive testing at Logie has shown this fear to be groundless. In fact, all the data suggest the opposite as the passage of fish is enhanced during the summer when water levels are at their lowest because the flow is now chan-nelled through a 6 m gap rather than spread over 42 m.

CHAPTER 4

THE HOMECOMING

One of the aspects of the salmon's life history which impresses not only students of salmon biology but also the general public is the ability of the fish to return to the place of its birth. In some instances, fish return after a lapse of four years and from feeding grounds which may involve a journey of some 5000 km. A Scottish priest, Hector Boece, in his 'History of Scotland', published in Latin (Boece 1527) and translated into Scots by John Bellenden in 1536 as the *History and Croniklis of Scotland*, refers to adult salmon returning to the place where they were born:

'They have a fervent desire and appetite to return to the places where they were born. Because many of the Waters of Scotland are full of waterfalls, as soon as they come to a fall they leap. Those that are strong or leap well, get up through the fall, and return to the place where they are bred, and remain there until their breeding season.'

Marking experiments date back to the mid-1600s (Walton 1653, Russel 1864). They involve the mutilation of particular fins or the attachment of some identifiable material to the smolts when they leave fresh water and catching them again when they return to the same place, usually some six months later. However, much doubt surrounded these early experiments and it was not until the beginning of the 1900s that conclusive proof of the salmon's ability to return home became generally available. This proof resulted from tagging experiments on the R. Tay (Malloch 1910). Although some tagged fish were caught in the lower reaches of other rivers, sufficient time remained for these fish to have returned to their home rivers before spawning if they had not been caught.

4.1 Homing

The homing aspect of the salmon's life history has interested many biologists. In the first phase of this process, the fish return to the coastline. In the second phase, they move along the coast, making their entry to the natal river on the basis of local cues. Salmon may make landfall well away from their natal rivers. It has been suggested that along the east coast of Scotland, for example, the fish reach land at particular points, one of these being Montrose Bay (Menzies 1938a). This suggestion is supported by the results obtained by tagging and releasing returning fish caught in bag nets fishing in Montrose Bay (Pyefinch & Woodward 1955, Pyefinch & Shearer 1957, Shearer 1958, 1986a). These fish were recaptured in many adjacent rivers along the coast from the inner Moray Firth to just south of the R. Tweed.

The plume of fresh water issuing into the sea from a river is thought to be an important guide to the fish migrating into it (Huntsman 1934). Entry is often associated with the occurrence of freshets (Huntsman 1945, Alabaster 1970). Salmon swimming offshore and along the coast have been observed to swim in particular directions (Hawkins *et al.* 1979a). However, because their swimming speeds are low in relation to tidal current speeds their movements are deflected, and they often appear to be swimming with the current. When they enter a river, their behaviour must change so that they swim against the current. Chemical cues may initiate this change in behaviour, the actual orientation of the fish being brought about by a response to the water current. In 1880, Frank Buckland suggested that salmon were assisted by their power of smell to find their way in the ocean and also to find their parent river. Many other ideas have also been put forward to explain how they achieve this, e.g. that they have the ability to detect temperature, current, and the amount and direction of light. The idea supported by most people is based on the results of a series of experiments by Hasler & Wisby (1951), and by Hasler (1954). These experiments showed, apparently conclusively, that salmon could detect individual stream odours provided that their olfactory organ was intact and functioning. However, if the olfactory organ was destroyed, salmon immediately became unable to return to their home river. The importance of olfactory cues to salmon migrating in the Baltic Sea has also been clearly demonstrated by Bertmar & Tofr (1969). Anosmic fish (deprived of their sense of smell) did not home, while control fish did. However, a recent study in Norway has suggested that the olfactory sense is not essential for smolt navigation through rivers and lochs (Døving *et al.* 1984). Recently, deposits of magnetite have been found in the bodies of salmon, particularly in the head and lateral line regions (Moore *et al.* 1990). Salmon may use these deposits for magnetic orientation.

The memory of the smell of the home river is not imprinted during the early life of the salmon but during the last few days before it enters the sea. This was neatly demonstrated by comparing the behaviour of wild smolts from the R. North Esk with that of an equivalent number of hatchery-reared smolts. These smolts were the progeny of spawners from the R. Conon and reared in water from the Inverness-shire R. Garry. All the smolts were similarly tagged before both batches were simultaneously released into the R. North Esk just above the head of the tide at the mid-point of the smolt migration in that river. Tagged fish from both release batches were recaptured during the following three fishing seasons mainly around and in the estuary of the R. North Esk, but not in north or west Scotland. The pattern of recapture sites was identical in the two groups of fish. However, the proportion of hatchery-reared smolts recaptured as adults was smaller than that of the wild smolts.

The imprinting hypothesis of Hasler & Wisby (1951) was extended by Harden Jones (1968) who suggested that salmon were sequentially imprinted during the

LIVERPOOL JOHN MOORES UNIVERSITY
LEARNING SERVICES

smolt migration. He proposed that salmon have to experience the outward migration as smolts and that sequential imprinting of the migratory smolts and post-smolts was necessary for their successful return as adults. The results of recent Norwegian experiments support this idea because smolts transported and released in the sea, and thus deprived of parts of the smolt migration, failed to return to their river of origin while fish from the same batch released in fresh water returned to the river of release (Hansen *et al.* 1988).

4.2 The time of return and progress upstream

In the United Kingdom, salmon may enter rivers in every month of the year. They not only enter the larger rivers such as the Shin, Ness, Spey, Dee, North Esk, Tay, Tweed, Avon and Wye but also smaller systems like the Naver, Thurso, Helmsdale, Brora, and Lune, provided there is sufficient flow of water in the rivers and that the temperature difference between sea and river is not too great. Salmon are usually caught by rod and line on these rivers before winter ends and some anglers fishing Loch Tay, some 100 km from the sea, on the opening day of the season (15 January) are frequently rewarded with a 'fresh run' salmon for their efforts.

Rates of upstream progress varying between 0.18 and 0.74 km day^{-1} have been recorded by Hawkins & Smith (1986) and Laughton (1989) for salmon tracked in the Rivers Dee and Spey, mainly in the summer. However, there are several problems associated with measuring progress upstream. For example, it is not certain that the fish swim continuously between two points of measurement. Moreover, slack water associated with pools might produce misleadingly high calculated speeds for short distances. Nevertheless, to have reached Loch Tay by January suggests that these fish must have left the sea some weeks if not months previously, particularly as upstream progress in winter, when the ambient temperature is marginally above freezing, is almost certainly slower than that in the summer.

4.3 The pattern of return

Salmon do not return from the sea to fresh water in a haphazard pattern, but according to their river- and sea-age (Shearer 1984a, 1990). Although data from the R. North Esk have been chosen to illustrate this pattern, less comprehensive data from the other major east coast salmon rivers indicate that this pattern of return is not unique to the R. North Esk.

Between 1963 and 1988, the length, weight and age structure of the salmon population returning to the R. North Esk in each month during the fishing season (16 February−31 August) and the age composition (river- and sea-age) of those fish returning out of season have been examined. The samples were taken from the river net and coble fishery during the fishing season (Plate 4), and by electro-fishing Kinnaber Lade in the close season. The data are thought

to be comparable because no significant difference was found between similar
sets of information relating to catches taken by these two methods during the
fishing season.

In all years, the two main components of the catch had spent either one
(1SW) or two winters (2SW) in the sea; together, these components accounted
for more than 90% of the catch. Although in some years their respective
strengths were approximately equal, in others there were marked differences.
The proportion of the 1SW component in the combined catches of 1SW and
2SW fish ranged from 22 to 75%. A third group of fish which had spent three
or more winters in the sea made its greatest single contribution (12%) to
catches in 1968. Since 1968, apart from the early 1980s, there has been a
general decline in the proportion of 3 or more SW fish in the total catch, down
to a minimum value of less than 1% in 1988. In most years, small numbers of
4SW fish were caught but not more than 1% in any year (Table 4.1).

The 1SW fish component sometimes entered the fishery as early as May and
by July was always the dominant sea age group. Although 2SW fish were

Table 4.1 Percentage sea age composition of R. North Esk net and coble catch of
maiden salmon in 1963−88.

Year	Age (sea-winter)			
	1	2	3	4
1963	35.7	61.6	2.7	0.1
1964	46.3	49.9	3.8	0
1965	50.8	43.3	5.9	0
1966	35.4	57.3	7.3	<0.1
1967	36.1	53.0	10.6	0.2
1968	27.5	60.1	11.7	0.7
1969	68.4	23.0	8.5	<0.1
1970	56.1	39.7	4.0	0.2
1971	33.9	61.5	4.5	0.1
1972	53.8	38.6	7.3	0.3
1973	58.4	38.3	3.2	0.1
1974	55.2	42.7	2.0	0.1
1975	21.6	74.4	3.9	0.2
1976	50.9	46.4	2.5	0.1
1977	54.3	43.4	2.3	0
1978	49.3	48.6	2.1	0
1979	55.0	43.5	1.5	<0.1
1980	40.3	56.0	3.7	0
1981	22.8	68.9	8.3	0
1982	48.4	43.5	8.0	0
1983	45.9	48.2	6.0	0
1984	55.4	40.2	4.4	0
1985	45.8	52.0	2.2	0
1986	66.9	30.9	2.2	0.1
1987	58.8	39.0	2.1	0.1
1988	70.5	28.7	0.9	0

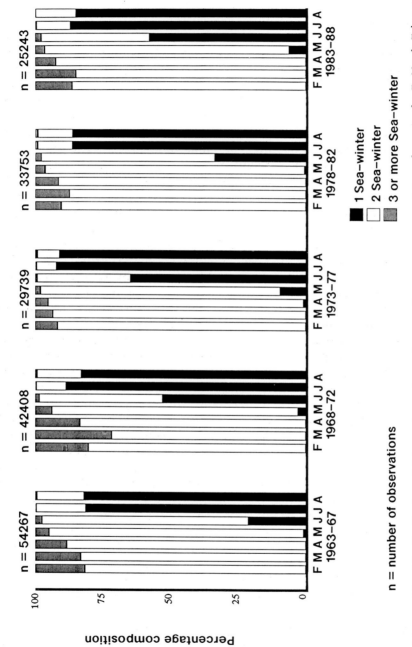

Fig. 4.1 Monthly distribution of the numbers of one-, two-, or three or more sea winter salmon in the R. North Esk net and coble catch in 1963–88, expressed as five-year means.

invariably the major component in catches from February to May or June but not later (Fig. 4.1), their importance in spring increased because of the decline of 3 or more SW fish. Very few three or more SW fish were caught in the second half of the fishing season.

Every year, fish representing four smolt age cohorts (i.e. fish which had spent 1, 2, 3 or 4 years in the river before going to sea) were present in the catches throughout the fishing season (Table 4.2). Irrespective of the number of years which these fish subsequently spent in the sea, the mean river age of the smolts in each sea age group declined throughout the fishing season (Figs 4.2 & 4.3). For example, the largest percentages of 1SW fish derived from four-year-old smolts were caught in May and from one-year-old smolts in August. Most 2SW fish arising from the three- and four-year-old smolt age-groups were caught in February. Most fish caught in August had spent only one or two years in the river before migrating to sea. The mean river age of 3 or more SW fish caught also declined from February to August. However, the smolt age distribution pattern for this group was less consistent because of small numbers caught (Fig. 4.4).

The demarcation between spring and summer salmon has been taken as 30 April because most salmon caught by the net fisheries up to 30 April had closed bands at the edge of their scales while fish caught from May had open bands. This classification, however, underestimates the number of rod-caught spring salmon, because some fish taken after 30 April (the proportion depending on the fishery) could have entered the river much earlier.

In the annual close season for netting from September to mid-February, two types of fish were caught. These are referred to as spawners and springers. Spawners include sexually mature fish which could spawn immediately and also immature fish which could not spawn until some months later in the same spawning season. On the other hand, the springers could not spawn until the following spawning season, probably during late October or early November. Bright silver fish with sea lice still attached occur in both groups so that these characteristics were not necessarily indices of maturity. In fact, some 'fresh run' spawners, whose scales, when examined, indicated that they had not earlier been in another river, extruded milt or eggs on the slightest pressure. These results indicate that salmon have the capacity to mature in the sea and do not require to return to fresh water. However, any extruded eggs would not survive in salt water. In years when the autumn rainfall has been below average and when the river flows are at or below summer level, some spawners, particularly the larger fish entering the river in November, frequently spawn on the first available gravel, which in many rivers may be in brackish water. When these redds are excavated, the eggs are found to be dead.

The dominance of 1SW fish already noted in the net and coble catches taken in July and August continued into the September trap catches but by October in some periods the proportions of grilse and salmon were about equal (Fig. 4.5a).

Table 4.2 Percentage smolt age (in years) composition of R. North Esk net and coble catch of maiden salmon in 1963–88 by sea age.

Year	1 sea-winter				2 sea-winter				>3 sea-winter				Multi sea-winter combined			
	1	2	3	4	1	2	3	4	1	2	3	4	1	2	3	4
1963	2.5	70.2	26.9	0.4	2.0	63.8	33.0	1.2	7.5	65.6	26.9	0	2.2	63.9	32.7	1.2
1964	0.3	73.9	25.6	0.3	2.5	71.1	26.3	0.2	7.4	72.5	19.0	1.1	2.8	71.2	25.7	0.3
1965	0.5	45.8	52.1	1.7	0.3	62.2	37.2	0.3	6.9	77.3	14.7	1.0	1.1	64.0	34.5	0.4
1966	0.9	56.9	39.4	2.6	0.8	40.4	54.9	3.9	1.2	72.1	26.7	0	0.9	44.0	51.7	3.4
1967	2.0	62.5	34.8	0.7	2.6	61.6	33.5	2.4	0.3	68.6	30.8	0.3	2.2	62.8	33.0	2.0
1968	3.1	60.7	34.5	1.8	0.9	51.5	45.0	2.6	1.1	70.5	27.5	0.8	0.9	54.7	42.0	2.3
1969	5.1	84.0	10.4	0.5	2.5	63.1	31.6	2.8	0	83.9	16.1	0	1.8	68.7	27.4	2.1
1970	1.3	69.5	28.7	0.5	1.7	72.3	23.6	2.4	2.6	66.1	30.0	1.3	1.8	71.7	24.2	2.3
1971	4.7	70.4	24.5	0.4	2.3	72.5	24.5	0.7	3.2	85.8	10.9	0	2.4	73.4	23.6	0.6
1972	6.4	60.1	33.0	0.5	3.5	66.4	29.6	0.5	2.1	77.3	20.6	0	3.3	68.2	28.1	0.4
1973	10.4	64.8	23.9	0.9	6.0	59.6	33.6	0.8	3.7	77.6	18.7	0	5.8	61.0	32.5	0.7
1974	2.8	78.7	17.7	0.7	6.8	63.9	28.1	1.1	6.0	52.0	42.0	0	6.8	63.4	28.8	1.0
1975	4.6	71.3	23.6	0.4	1.7	75.1	22.4	0.7	12.6	65.8	21.6	0	2.3	74.6	22.4	0.7
1976	10.9	74.0	15.0	0.2	6.0	64.0	29.5	0.5	2.7	86.4	9.5	1.4	5.8	65.2	28.5	0.5
1977	8.1	59.6	31.9	0.4	7.5	68.3	23.2	1.1	21.1	66.7	12.2	0	8.2	68.2	22.6	1.0
1978	7.9	73.0	18.8	0.3	7.6	60.0	32.2	0.2	12.3	74.2	13.5	0	7.8	60.6	31.4	0.2
1979	5.9	78.1	15.7	0.3	3.3	61.2	33.8	1.6	10.6	54.5	35.0	0	3.6	61.0	33.9	1.5
1980	4.4	72.6	22.8	0.3	7.7	65.3	26.7	0.2	14.4	70.7	14.8	0	8.1	65.7	26.0	0.2
1981	1.5	61.7	36.8	0	1.2	67.5	29.6	1.7	11.5	63.2	23.5	1.8	2.3	67.0	28.9	1.7
1982	5.4	68.8	25.2	0.7	0.6	50.3	47.4	1.7	0	76.3	22.2	1.5	0.5	54.3	43.4	1.6
1983	3.3	71.2	24.9	0.7	1.1	30.6	60.5	7.8	0	57.6	40.4	2.0	1.0	33.6	58.3	7.1
1984	2.7	58.5	36.8	2.0	1.8	55.4	37.1	5.7	0	62.0	38.0	0	1.7	56.0	37.2	5.1
1985	1.1	74.3	22.9	1.7	3.1	50.8	41.0	5.0	5.3	59.6	35.1	0	3.2	51.2	40.8	4.8
1986	5.2	66.7	27.2	1.0	2.3	61.7	32.5	3.5	0	62.5	28.8	8.7	2.2	61.7	32.2	3.9
1987	2.5	68.0	28.6	0.9	5.6	56.5	34.7	3.2	0	86.4	13.6	0	5.3	58.1	33.6	3.1
1988	1.8	51.1	45.3	1.7	2.1	73.5	24.3	0.2	7.0	63.2	29.8	0	2.2	73.2	24.4	0.2

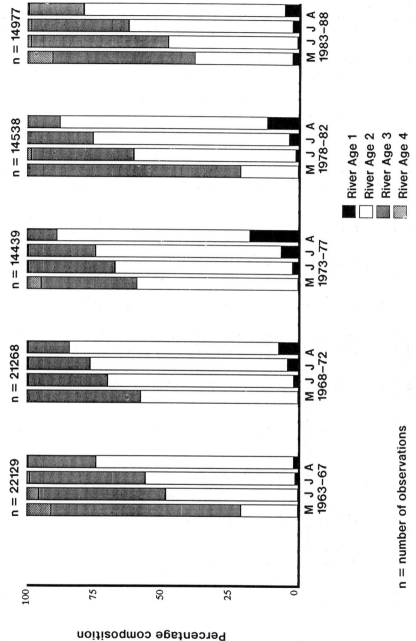

n = number of observations

Fig. 4.2 Monthly smolt age composition of one sea-winter salmon caught by net and coble in the R. North Esk in 1963–88, expressed as five-year means.

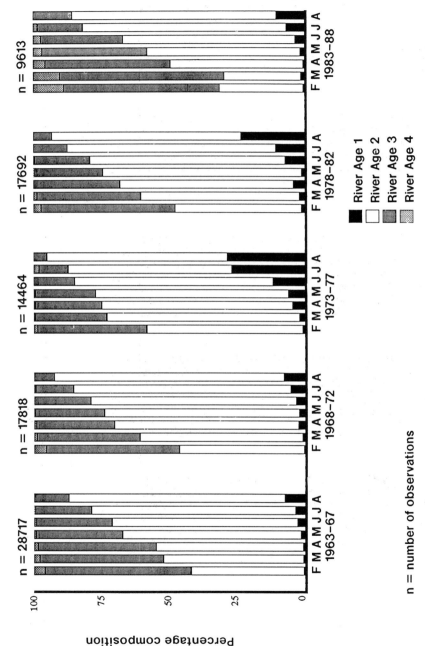

n = number of observations

Fig. 4.3 Monthly smolt age composition of two sea-winter salmon caught by net and coble in the R. North Esk in 1963–88, expressed as five-year means.

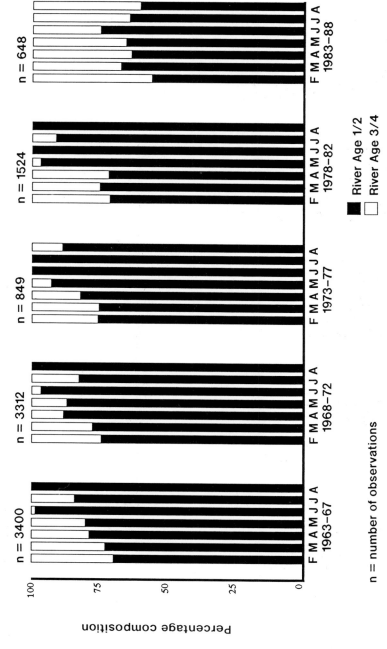

Fig. 4.4 Monthly smolt age composition of three or more sea-winter salmon caught by net and coble in the R. North Esk in 1963–88, expressed as five-year means.

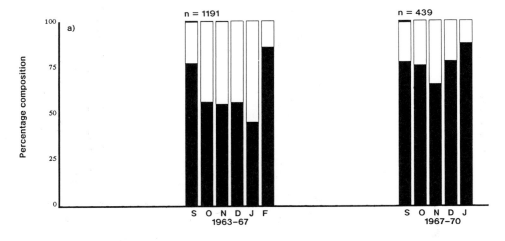

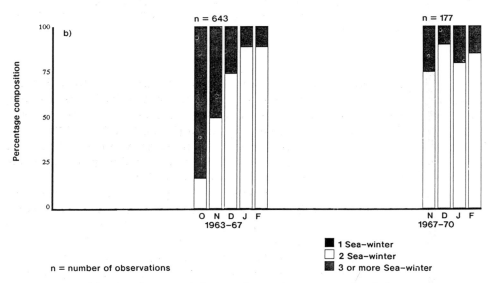

Fig. 4.5 Monthly distribution of the numbers of one-, two-, and three or more sea-winter spawners (a) and springers (b) in trap catches between 1963 and 1970, expressed as four (1963/64−1966/67) and three (1967/68 to 1969/70) annual close season means respectively.

Two sea age groups (3SW and 2SW) were also present in the catches of springers. Fish entering their third sea-winter usually dominated catches in October but from December to at least February, when trapping ceased, the dominant sea age group was a year younger (Fig. 4.5b). All four river age groups (1, 2, 3 and 4 years) were identified in both these groups of fish, but

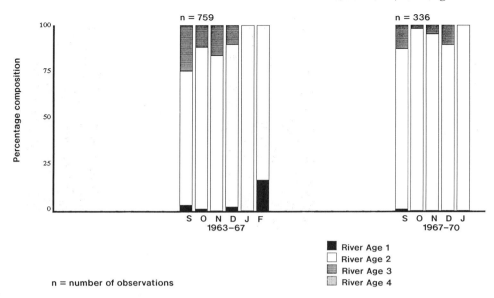

Fig. 4.6 Monthly smolt age composition of one sea-winter spawners in trap catches between 1963 and 1970 expressed as four (1963/64−1966/67) and three (1967/68−1969/70) annual close season means respectively.

most had spent two or three years in fresh water before smolting. A small number of 1SW spawners had spent four years in fresh water before smolting but no 2SW spawner had spent more than three years, and less than 10% only one year, in fresh water before migrating to sea (Figs 4.6, 4.7a & 4.8).

Most springers had spent two years in fresh water before smolting, and more had spent three or four years in fresh water than the spawners. None had migrated to sea as a one-year-old smolt (Figs 4.7b & 4.8).

In 1966, a typical year, 1SW fish were first caught in May. Their numbers peaked in July and they continued to make a substantial contribution (>30%) to catches until January 1967; 2SW fish first appeared in November 1965 and from December 1965 until June 1966 they formed 90% of the catch − a figure which then declined to about 25% of the total monthly catch. During the autumn, numbers of 2SW fish again increased coincident with the entry of autumn fish into the river. Only in one month (November 1965) did 3SW fish dominate catches (60%), and by May they had virtually disappeared. Between November 1965 and January 1967, the mean smolt ages of the 1SW and 2SW fish decreased (Figs 4.9 & 4.10).

Previous spawners also returned from the sea according to a fixed pattern. They returned on approximately the same date on each spawning migration. For example, fish which initially spawned as grilse, summer salmon or autumn salmon spent less than 12 months in the sea between subsequent visits to fresh water. Spring fish normally only returned to spawn in alternate years. Previous spawners contributed less than 5% to the spawning stock of east

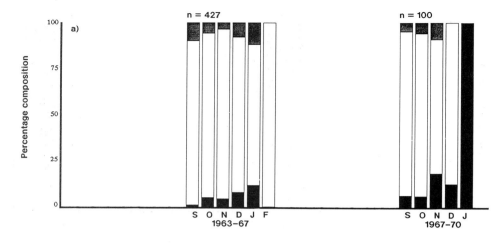

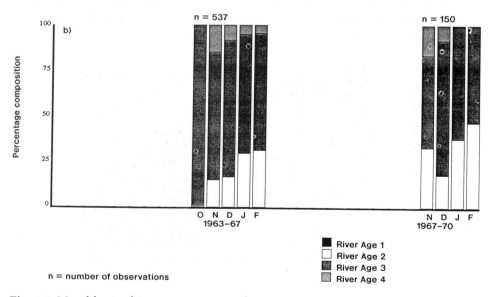

River Age 1
River Age 2
River Age 3
River Age 4

n = number of observations

Fig. 4.7 Monthly smolt age composition of two sea-winter spawners (a) and springers (b) in trap catches between 1963 and 1970 expressed as four (1963/64−1966/67) and three (1967/68−1969/70) annual close season means respectively.

coast rivers and about 10% in west coast rivers. Most (>95%) previous spawners were female.

Although particular sea and river age groups dominated catches in particular months, each age group contained fish from more than one spawning stock.

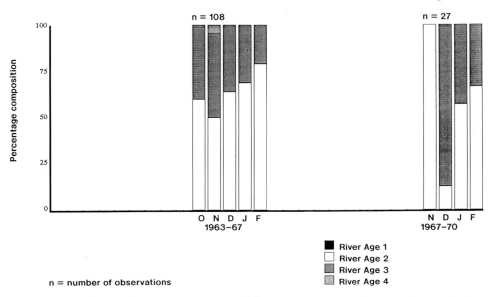

Fig. 4.8 Monthly smolt age composition of three or more sea-winter springers in trap catches between 1963 and 1970 expressed as four (1963/64−1966/67) and three (1967/68−1969/70) annual close season means respectively.

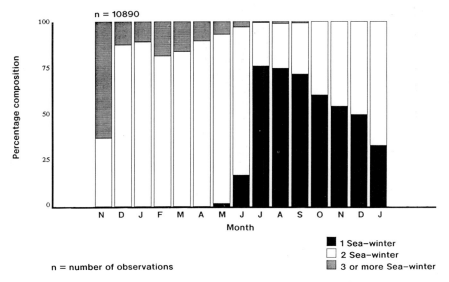

Fig. 4.9 Monthly sea age composition of the 1966 spawning stock.

The incoming stock was never entirely dependent on the progeny from a single spawning season, and fish from a given spawning season returned over a long period. From the 1966 spawning season, for example, maiden adult fish returned from three to eight years later (up to 1975), as a result of the various

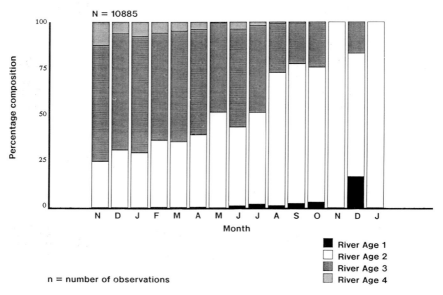

Fig. 4.10 Monthly smolt age composition of the 1966 spawning stock.

combinations of river and sea age. From each cohort of smolts, fish return over a four-year period. Hence, there is an in-built limited insurance against the extinction of stocks if, in a particular year, disaster should strike and kill all the juveniles in fresh water or all the adults at sea. Rivers whose stocks produce a smaller range of river and sea age groups than the R. North Esk will be less well protected.

Of the fish spawning in any year, 95% enter the river over a 12-month period from 1 December. The season of migration largely depends on the age composition of the returning stock and whether it contains the full range of age groups. In grilse-only rivers, immigration may last no more than six months. In Canada and Norway, unfavourable river conditions may prevent springers from entering fresh water.

In every month of the years 1981−88 (the years for which quantitative data are available), some fish migrated into the R. North Esk. There was a big variation in numbers between months, ranging from 2−3924 fish. In five of the nine spawning years, most fish crossed the counter at Logie in September (Fig. 4.11), and 24−69% of each annual spawning stock migrated upstream over Logie after the fishing season had ended on 31 August.

The age composition of the catches taken on the various beats on the major Scottish east-coast salmon rivers shows a general overall pattern. The ratio of grilse to salmon declines as the distance from the sea increases (i.e. more salmon on the upper beats) (Fig. 4.12). A change in the sea age of the returning adults towards grilse, which mostly enter fresh water after mid-summer, may dramatically reduce the availability of any fish, but especially salmon, on the upper beats. In these circumstances, the overall decline in the 3SW and 4SW group also means that fewer big fish in the 15−20 kg weight

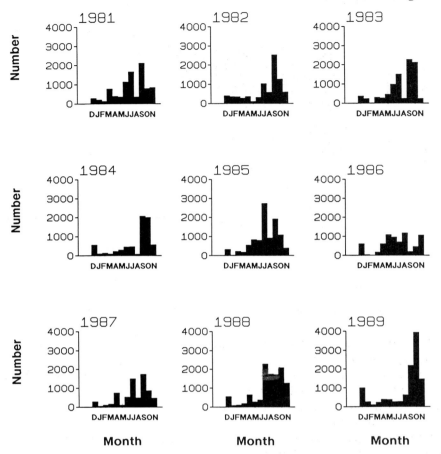

Fig. 4.11 Numbers of potential spawners crossing the Logie fish counter each month in 1981−9.

range are available to be caught.

Fish which spent 1−4 years in the river before going to sea as smolts were identified in the rod and line catches, but the various age groups of smolts were not equally distributed along the length of the river. The mean river age of the rod-caught fish increased with distance upstream (Fig. 4.13). Because the 1SW and 2SW fish which spent the shortest time in the river before smolting enter fresh water late in the year, smolt age is important to rod fisheries in general, and especially to those furthest from the sea (Shearer 1988a).

Although the catch of spring fish declined and the grilse catch increased in 1964−88, the numbers of grilse spawning in the largest spring fish spawning tributary of the R. North Esk, the Water of Mark, did not increase significantly. Nor was there any change towards younger fish in the age composition of the juveniles, and the grilse were below the average length for the R. North Esk, suggesting that they had entered the river before the mid-point of the grilse run.

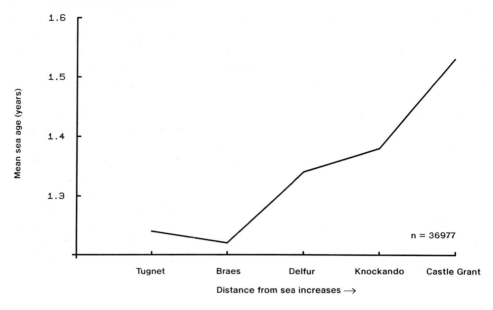

Fig. 4.12 Mean sea age of catches taken on various fisheries along the R. Spey.

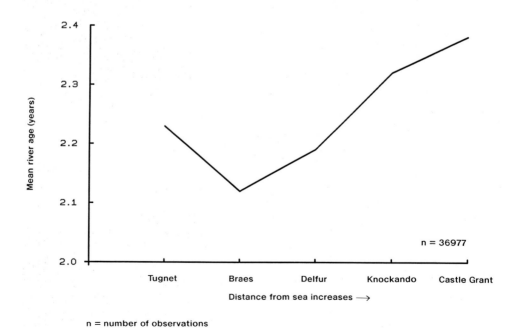

Fig. 4.13 Mean river age of catches taken on various fisheries along the R. Spey.

These data show that any decline in the early running MSW component of the stock, particularly in those fish which have spent three or more years in fresh water before migrating to the sea as smolts, is likely to depress the number of fish available to anglers fishing the upper reaches much more than on beats nearer the sea. At these upstream beats, an increase in the numbers of returning grilse does not fully compensate for this loss.

Data based on the great mass of circumstantial evidence, mostly obtained from ageing adult fish entering the river at different times of the year and juvenile fish in a wide range of spawning burns over many years, give rise to the belief that the parents of early running fish are generally also early running and spawn in the head-water tributaries of the main river. Further support for this idea has come in recent years from the results of radio tracking salmon ascending the Rivers Spey, Dee and Tay (Plates 14 and 15). The results of all these experiments showed that earlier entrants spawned further upstream than those entering later (Hawkins & Smith 1986, Laughton 1989, Webb 1989). This radio tracking also confirmed the earlier observations of Dunkley & Shearer (1982) that most upstream migration occurred at night,

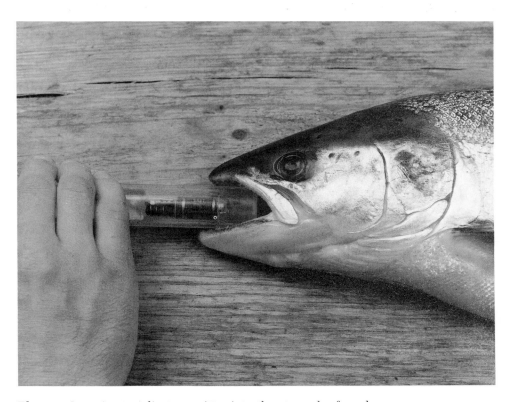

Plate 14 Inserting a radio transmitter into the stomach of a salmon.

Plate 15 Tracking radio tagged salmon.

from dusk to dawn. Movements in daylight were most frequent in spates except at spawning time when movement occurred at any time, day or night, irrespective of any change in water level.

4.4 Summary

Homing salmon return to the coastline, and then reorientate themselves, on the basis of local cues, so that they enter their natal river. The memory of the smell of the home river is imprinted during the last few days before the fish enter the sea. Salmon must experience the outward migration as smolts for their successful return to a particular stream as adults.

Most salmon return after spending either one (grilse) or two winters in the sea. Grilse are found in most rivers and enter fresh water as early as May. By July, grilse form the dominant sea age group in most rivers. Salmon, by definition older fish and including a relatively small proportion of 3 or more SW fish, are infrequent in many Scottish west-coast rivers. Salmon begin entering the big east-coast rivers in October. The few 3 or more SW fish enter before June of the following year but 2SW fish continue to arrive for about 12 months. Some fish entering in October in one year may stay and not spawn until the following October. Fish which have spawned previously contribute less than 5% to the spawning stock of most east coast rivers, and rather more (*c* 10%) in west-coast rivers.

Within each sea age-group, the fish derived from the oldest smolt age group return earliest. Particular sea age groups may dominate catches in some months, but there is no month when the incoming stock depends entirely on the progeny from a single spawning stock.

Different angling beats along the length of each river tend to exploit different stocks of fish and they may differ from the stocks exploited by the net and coble fishery.

Although most 2SW fish are heavier than 1SW fish caught in the same month, towards the end of the season grilse are on average heavier than 2SW fish caught in the early spring.

In the R. North Esk, most grilse are male, and most 2SW fish are female. However, in 'grilse only' rivers, the proportions of male and female grilse are about equal. The parents of early running fish are also generally early running. They are thought to spawn in the headwaters of the main stem and its upper tributaries. Most grilse spawn in, and are available for capture in, the lower reaches and its tributaries.

CHAPTER 5

THE FRESHWATER PHASE

In the British Isles, this phase of the life-history lasts one to four years. However, the length of time that salmon parr spend in fresh water varies with the location of the river. In many southern rivers draining into the Atlantic Ocean, Irish Sea or English Channel, most smolts are yearlings. For example, in the Hampshire R. Avon over 90% of the smolts are one or more years old. On the other hand, the majority of the smolts leaving Scottish rivers each spring are either two or three years old and the proportion of one-year-old smolts seldom exceeds 10%. The age of smolts increases towards the northern limit of the salmon's range, and smolts aged seven years are not uncommon in northern Norway. In some rivers draining from glaciers, the mean smolt age can be six years (Jensen & Johnson 1986). In any particular river, the average smolt age is correlated with the number of days in each year on which the temperature reaches or exceeds 7°C (Symons 1979).

5.1 Fecundity

In studies of the fecundity of salmon in Scottish rivers, Pope *et al.* (1961) and Shearer (1972) discovered a relationship for each river between the length of a female salmon and the number of eggs produced by that fish. We pointed out that there were statistically significant variations in this relationship between different rivers and between fish from different tributaries of the same river. The average egg counts corresponding to a fish of length 70 cm (the average length of all salmon handled in both studies) on each of the seven rivers studied were: R. North Esk, 4600; R. Lyon, 4943; R. Blackwater, 5117; R. Garry (Inverness-shire), 5370; R. Dee, 5495; R. Conon, 5572 and R. Meig, 6067. The Rivers Blackwater and Meig are tributaries of the R. Conon.

As examples, the length fecundity relationships for salmon from the Rivers North Esk and Meig were respectively of the form:

$$\log_{10} N = 3.0668 \log_{10} L - 1.9955 \quad \text{and}$$
$$\log_{10} N = 2.3345 \log_{10} L - 0.5333$$

where N = number of eggs and L = fork length, in cm.

When great accuracy is not essential in estimating potential egg deposition, it may be sufficient to employ the general equation:

$$\log_{10} N = 2.3345 \log_{10} L - 0.582$$

where N = number of eggs and L = fork length, in cm.

However, when the earlier experiment was repeated in 1987 and 1988, the average egg count corresponding to a fish length of 70 cm was 7706 and the fecundity relationship for salmon from the R. North Esk based on a sample of 60 fish ranging in length from 59.5 to 92 cm was of the form:

$$N = 297.6 L - 13126.3$$

where N = number of eggs and L = fork length, in cm.

Because both length and number of eggs were normally distributed, no log transformation was necessary for these data.

All the fish used in the two earlier fecundity experiments had been caught just prior to spawning in the upper reaches of the respective main river or tributary. An examination of their scales revealed that they had been in fresh water for many months. In the more recent experiments, the fish were caught in the lowermost 5 km of the R. North Esk in November and had only relatively recently entered that river from the sea. The increase found in 1987 and 1988 in the average egg count of R. North Esk salmon corresponding to a fish length of 70 cm may have resulted from differences in the time of entry into the river. The more recent experiments used fish which did not require to maintain themselves for many months and also swim upstream (*c* 50 km) on energy realized from their food reserves. Consequently, they had more resources to convert into eggs. In addition, the mean egg size of the later running fish was larger than that of fish of equivalent length which entered fresh water in the first half of the year.

The remaining data required to estimate the potential egg deposition in any river system are the number and lengths of the females in the spawning population. On the R. North Esk, this information was obtained from the automatic fish counter at Logie and from the trap in the lade at Kinnaber Mill (Plate 11 in the colour section). The biological data obtained from the trap catches were supplemented by sampling the in-river net and coble catch throughout the year.

5.2 Egg deposition

In the years for which counter data are available (1981–89), the potential egg deposition in the R. North Esk varied between 14 and 29 million eggs, which is equivalent to densities of between 6.6 and 13.3 m^{-2} of fry-rearing habitat. These figures are significantly higher than the egg deposition of 2.4 m^{-2} recommended for Newfoundland rivers, the value which is also used by the Canadians to assess the number of female spawners which are required to stock a stream to its maximum carrying capacity (Marshall 1984, Randall 1984, Chadwick 1985). However, part of these differences may arise because the criteria used to describe fry-rearing habitat may differ between workers.

Elson & Tuomi (1975) found that the egg deposition in the R. Foyle was 1.68 m^{-2} while Buck & Hay (1984) recorded a maximum egg deposition of 3.4 m^{-2} in the Girnock Burn which is a tributary of the Aberdeenshire R. Dee and some 10 km distant from the source of the R. North Esk.

5.3 The juveniles

5.3.1 Density

Young salmon have been described as territorial, but the idea that juveniles defend exclusive territories in the wild is becoming increasingly doubtful. They may show aggressive behaviour and compete with one another but the social structure individuals adopt is not clear. Water depth, water velocity, overhead cover and substrate size are important in determining distribution. The physical environment in a stream is highly heterogeneous, and there are marked differences in habitat quality between sites only a short distance apart. In the Girnock Burn, young salmon are spatially aggregated (clumped) at all juvenile densities, and the home ranges of individuals overlap (Garcia de Leaniz 1990).

Mean parr densities vary between rivers and between sites in the same river system. An electro-fishing survey in 1988 involving sampling at 129 sites within the catchment of the R. Tweed showed that the densities of both fry (0–305 100 m^{-2}) and parr (0–136/100 m^{-2}) varied widely (Gardiner 1989). High densities of fry and parr generally occurred at the same sites. At some locations, the lack of fish may have been due to the native fish not having been able to ascend the tributary to spawn, perhaps because of permanent barriers such as falls or weirs or as a result of unfavourable flow regimes or periodic pollution. Comparable data collected in the previous year, although limited, suggested that densities of fish were similar at the same site at least over short periods. Salmon parr have very definite preferences and significantly select some sites in a stream and avoid others. In the summer, salmon parr choose sites that maximize food intake and minimize energy expenditure and in winter they seek shelter and cover. When the water temperature drops below 5°C they hide within coarse (>20 cm) substrates in riffles. In general, salmon parr select deeper waters, coarser substrates and faster currents than fry. Until we know what all their preferences are, it will inevitably be difficult to understand why densities are different at different locations.

At a site in the R. Tweed watershed measuring 525 m long and, on average, 3.25 m wide, the density of juvenile salmon was increased by the transfer of fry from a nearby location. Initially, the enhanced density of fish declined from the target density of 210 fish 100 m^{-2} to *c* 21–130 fish 100 m^{-2}. However, when the site was revisited some three months later, the density of juveniles remained amongst the highest recorded instead of the lowest in the R. Tweed. Consequently, the rearing potential of this site had been primarily under-utilized. There may be other similar sites in the R. Tweed where access to spawning fish is not the limiting factor. From this, juvenile densities, apart from indicating the presence or absence of fry and parr, are of limited value as quantitative measures of the level of utilization of stream areas or as indices of the number of spawners. They are of even less value in determining the annual smolt production in the absence of any measure of the amount of wetted area, which can change markedly between years.

The population densities of those parr (9 cm or greater fork length) in headwater tributaries of the R. North Esk which were likely to migrate as smolts in the following spring were estimated in the summer months in the 1970s and in the 1980s. In the Water of Mark, mean densities (number of parr 100 m^{-2}) in the 1970s ranged between 12 and 32 and in the 1980s between 17 and 30. In the Effock Burn, the comparable data were 14−27 in the 1970s and 16−28 in the 1980s, showing no change. However, the amount of habitable nursery area declined, particularly in the Water of Mark, and the total production of parr must also have declined. Although densities of parr in the tributaries Mark and Effock were similar, parr of 9 cm or more in length in the Water of Mark were, on average, significantly older than the corresponding length group of fish in the Effock Burn (2.1 compared with 1.7 years). All the evidence suggests that in each of these burns the fish are genetically distinct. Juvenile density data for other rivers are shown in Table 5.1.

5.3.2 *Dispersal*

The number of fish produced each year may be fixed very early in juvenile life at a time when the young salmon have little capacity for active swimming. The chances of survival at this stage depend largely on the site chosen by the parent female to construct her redd and deposit her eggs. However, smolt production depends not only on the number of redds dug but also on dispersal, i.e. the distance between the furthest upstream and furthest downstream redds. Downstream dispersal from a redd is more efficient than upstream dispersal. For example, over a period of four weeks, the young fish emerging from a redd spread out to occupy an area ranging from 166 m above the redd to 743 m downstream (Anon 1989a).

Survival from a particular brood year may also depend on the density of parr already present in the stream. In the Girnock Burn, the elimination of juveniles belonging to a particular year class caused the remaining year classes to smoltify at a younger age. This is because the carrying capacity of a nursery area may be limited by biomass rather than absolute numbers of fish.

The practical implication of these results is that the maximum stocking of a tributary can occur only if the returning females have free access to its headwaters. This entails minimizing physical obstructions or circumventing them with fish passes. In addition, sufficient water flow at spawning time to allow the females to enter the spawning tributaries and then disperse throughout their length is also most important, as is the maintenance of the wetted area during periods of drought.

Table 5.1 Densities of salmon fry and parr in Ireland, Scotland, England and Wales obtained from electro-fishing surveys.

Source	Mean salmon densities per 100 m^2 (maxima in parentheses)	
	0+	1+
Northern Ireland		
R. Bush	32.5 (231.3)	7.1 (24.5)
R. Foyle and tributaries	13.9 (49.0)	8.2 (46.0)
L. Neagh catchment	56.2 (236.0)	29.8 (92.0)
R. Camowen (pre-drainage)	78.3 (169.8)	12.0 (32.8)
L. Erne tributaries	20.6 (71.0)	11.5 (39.0)
Republic of Ireland		
Corrib system	153.8 (730.0)	20.6 (121.0)
R. Erriff	42.9 (68.0)	16.6 (25.0)
R. Suir	21.6 (77.8)	16.6 (46.2)
L. Currane catchment	30.3 (78.0)	8.4 (15.0)
Scotland		
R. Tummel	56.0 (107.0)	27.8 (36.0)
Shelligan burn	206.2 (309.0)	33.4 (97.0)
Wales		
R. Wye	94.6 (197.2)	9.5 (20.2)
England		
R. Exe tributaries	15.6 (30.1)	3.3 (6.3)
R. Lune	27.6 (391.6)	7.5 (70.4)
R. Esk, Yorkshire	3.6 (15.3)	2.7 (8.8)

Source: Mills (1989)

5.3.3 Homing

Salmon parr tended to stay in the same places in the Girnock Burn. Some parr which had been displaced by up to 240 m were able to return to their original site (homing). However, homing depended largely on the direction of displacement because fish displaced downstream homed much better than fish displaced upstream. Fish that were displaced downstream moved closer and closer to their homes and some succeeded in homing precisely while those that were displaced upstream generally strayed further and further away. Although intact parr may home better than fish deprived of their sense of smell (anosmic fish), vision may be enough to allow a parr to recognize its home area.

Homing is perhaps related to habitat quality — good sites are worth retaining by the fish, bad ones are not. Interaction with other fish may also be important for some displaced individuals: 'if a place is occupied, you move; if it isn't — and your previous home was not that good anyway — you stay'. If this is true, then larger fish occupying the prime homes would have a lot to lose and should home better following displacement than smaller fish which occupy poorer sites and for which the choice may simply be a 'best of bad' situation. In fact, homing success is found to be size-dependent, as larger fish were found to home better than smaller ones, other factors being equal (Anon 1989b). It could be that more sites are suitable for small fish than for large fish and as a result there is less incentive for the small fish to home.

5.3.4 Mortality

Causes of mortality in salmon fry have been studied since the mid 1960s in the Shelligan Burn, a small, productive tributary of the R. Almond, which is itself a tributary of the R. Tay. During the first year of life, when mortality rates are highest, an average of 1% of the Shelligan fry population dies each day. The rate at which these young fish die is partly a function of their density on the bed of the burn, reflecting the fact that in a fully stocked tributary, food and space rather than numbers of eggs set the main limits to fish production. The studies centred on two possible causes of this mortality, starvation and predation.

5.3.4.1 Starvation and predation — and disease

Results showed that salmon fry and parr both defended home ranges. The size of each home range depended upon the nature of the substratum, the area that a fish can patrol, and the feeding capacity of the stream bed. Some fish were squeezed into smaller territories and were finally excluded from the stream bed. Starvation could, therefore, be an important cause of mortality. During periods of starvation, fish were also likely to be more prone to disease because the lack of protein could suppress the production of enzymes, particularly for detoxification and immunological activity, so lowering resistance to disease.

The effect of predation was less clear. Judged merely on the basis of the food found in trout stomachs, 40% of the salmonid fry losses could be attributed to predation. On the other hand, when predators — including birds, salmon parr and trout — were excluded from a section of the stream, loss rates within an enclosure were no different from those occurring in the adjacent section of unprotected burn. This suggests either that predators concentrate on fish dying from starvation and disease, or that the fish remaining after predation compensate for the loss of their peers by growing and surviving better in an environment where there is less competition for food and space (Anon 1989b).

5.3.5 *Maturity*

It has been known for many years that, in addition to adults of both sexes which have returned from the sea, Atlantic salmon populations in rivers may also include considerable numbers of mature male parr which have not smolted. These fish stay close to adult females during redd construction, and at spawning release their milt at the same time as the adult male.

In 1987, using genetic markers, it was estimated that the proportion of eggs fertilized by male parr in individual redds in the Girnock Burn ranged between 1.4 and 25.2%. These are minimal figures, but the mean (12.3%) agrees with the results obtained under artificial conditions. High reproductive success of mature male parr has important genetic implications. The presence of mature male parr in a spawning area increases the number of individuals participating in reproduction, an important factor in the maintenance of genetic variability within a population (Anon 1989b).

An interesting result from the electro-fishing surveys in the tributaries Mark and Effock was that following the first year's marking of parr in these tributaries, fish which had been marked during the previous year were recaptured in each year's fishing. Some of these parr were caught in three successive years, attaining lengths greater than 17 cm (fork length). These males were maturing to spawn on each occasion that they were caught.

Maturing female parr are much less common and are mostly found in lochs which have been stocked and in impoundments where the outlet may be difficult to find.

5.3.6 *Food*

Juvenile salmon feed by ingesting whole animals, mainly invertebrates but occasionally small fish. The stomach contents often contain organisms of benthic origin taken either from the substrate or drifting in the current, organisms normally closely attached to stones, animals moving away from the bottom to the surface of the water at times of emergence, animals living on the water surface, and fauna of terrestrial origin attracted to the water surface, or washed or blown into the stream. They also take salmon eggs.

In a study of the food of young salmon in the Shelligan Burn and the R. Almond (tributaries of the R. Tay) Egglishaw (1967) showed that although the range of organisms was wide those which occurred most frequently in large numbers in the stomachs were the juveniles of *Hydroptila* spp (caddis fly) *Simulium* spp (black fly), *Baetis* spp (May fly), and *Leuctra* spp (stone fly). Other workers including Allen (1941) on the R. Thurso and Frost (1950) on the R. Forss have reported that the juvenile salmon in these rivers ate a similar diet.

Differences have been found in the diet of salmon taken in different habitats. For example, in pools, they fed less on *Baetis* spp and more on Chironomid pupae (midge) (Egglishaw 1967). There was also less food in the stomachs of salmon during the colder months of the year, November to February, than in the warmer months. In addition, more fish were caught with empty stomachs in the winter compared with the summer.

5.3.7 Migration

5.3.7.1 Parr

The migration of parr, unlike that of smolts, is not restricted to a fixed period. However, it tends to be more intensive during the late autumn. Another difference from the smolt migration is that the migration of parr is not uni-directional. However, Garcia de Leaniz (1990) showed that the movement of immature parr in the Girnock Burn in the autumn was downstream. The movement does not appear to be restricted to any one tributary or section of river. Parr marked in the late summer in upper and lower tributaries of the R. North Esk have been caught in Kinnaber Mill trap in both the autumn of the same year and in the spring of the following year. Those caught in the spring of the following year had smolted in the interval.

In most years, the greatest numbers of parr were caught in September in Kinnaber Mill trap irrespective of the direction of travel. The annual numbers of parr migrating upstream and downstream through Kinnaber Mill trap in 1974−88 are shown in Fig. 5.1. Most parr were one year old or more when caught and were, therefore, potential two-year-old smolts. Precocious male

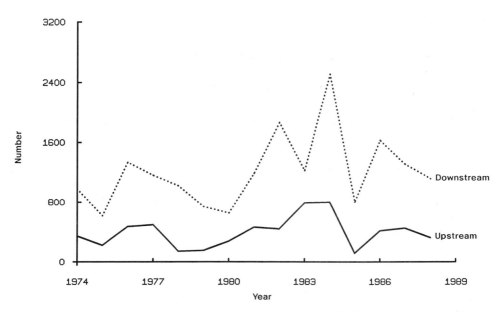

Fig. 5.1 Annual numbers of parr migrating upstream and downstream through the Kinnaber Mill trap in 1974−88.

parr appeared in both groups. The mean body lengths of the upstream and downstream migrating parr each year were similar and not statistically different from the mean body lengths of smolts migrating in the spring. Because no parr were caught in routine sampling in the estuary of the river, it is assumed that these fish did not overwinter in the estuary although they are capable of surviving in this area by avoiding full strength (33‰) sea water. This assumption was confirmed each spring when a proportion of the parr (upstream and downstream migrants) tagged the previous autumn were caught in the trap. In order to have gained access into the trap, these parr must have ascended above Morphie Dyke and entered Kinnaber Lade at its exit from Morphie Dam. The implication is that these parr overwintered upstream of Morphie Dyke.

5.3.7.2 Smolt

The transformation from parr to smolt involves a number of morphological, physiological and behavioral changes which pre-adapt young salmon for life in the sea while they are still in fresh water. In spring, during transformation of the parr to the smolt stage, the fish becomes more streamlined; a subcutaneous deposit of guanin is laid down, concealing the parr markings (dark blotches along each side) and giving the fish a silver colour, while the pectoral and caudal fins turn black.

Smolts emigrate from the R. North Esk each year in April, May and June, the peak time varying between years. During the years 1964−88, the mid-migration period (the five-day period which included the day by which 50% of the smolts had migrated) was earliest in 1967 (1−5 May) and latest in 1968 (26−30 May). There was a significant correlation between the timing of the mid-migration period and the average maximum daily water temperature in January−April; the higher the temperature, the earlier the smolt run. In Norway, on the other hand, Jonsson & Ruud-Hansen (1985) found that the start of the migration was not triggered by a specific water temperature or by a specific number of degree-days, and that there was no significant correlation between smolt descent and water flow, turbidity or lunar cycle. In Norway, there was no difference in survival of smolts emigrating at different times within the run. This conclusion was based on the return rate of individually tagged smolts, within each age/length group.

5.3.8 Sex and age

Over the whole run, female fish predominate in the R. North Esk, comprising about 64% on average of the emigrating smolts. However, the sex ratio changed during the course of the smolt run and, by the end of the run in June, there were more males than females. In 1964−70, the proportion of smolts which were female was significantly greater than the proportion of parr which were female (55%) in the sample from Kinnaber Mill trap between August and March.

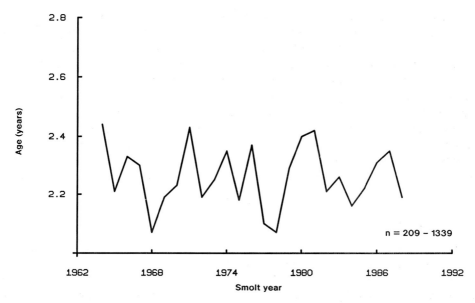

Fig. 5.2 Annual mean age of R. North Esk smolts in 1964—88.

The age composition of the smolts also changed throughout the run in the R. North Esk. Older smolts (three- and four-year-olds) emigrated earlier than younger ones (one-year-olds). The mean smolt age in April each year was greater than the mean age of smolts in May, which in turn was greater than the mean age of the smolts in June. Although no significant differences were found between months for any year, this pattern occurred from 1964 onwards (Dunkley 1986). The changes in mean smolt age, mean smolt length and sex ratios during the run were associated with the contribution being made by the different tributaries. Tributaries located in the headwaters — for example, the Water of Mark — produced relatively older and longer smolts, with a higher proportion of females and made their highest contribution early in the run. In 1964—88, there was no apparent change in the mean smolt age (Fig. 5.2). However, there was a steady decrease in the annual smolt length in the 1960s and 1970s and in the most recent years (Fig. 5.3), five-year rolling averages have been used to smooth the data and to show the underlying trend. Mean smolt length varied between 12.0 cm in 1985 and 13.0 cm in 1967. This trend seems to be incompatible with the change in the stock from salmon to grilse. With fewer MSW salmon spawning in upper tributaries, an increase in the body length of smolts would be expected. The decline in length in the 1960s to 1970s was not associated with the arrival of more spring salmon in that period.

Resulting directly from the low flows during the period of the 1990 smolt migration, the number of smolts caught in Kinnaber Mill trap was at least three times greater than that in the most recent years. In addition, there was a more equal monthly distribution of migrating smolts caught in the trap, and

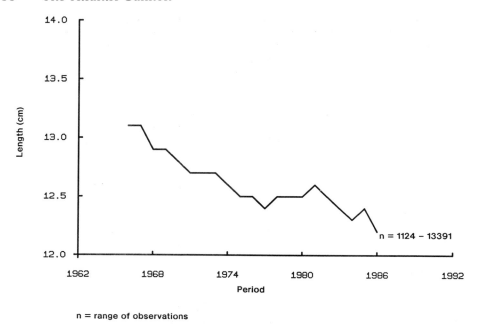

n = range of observations

Fig. 5.3 Annual mean length of R. North Esk smolts in 1964–88, expressed as five-year rolling averages.

the number sampled was considerably larger. Nevertheless, there were no major differences in the mean length (12.2 cm) of the smolts or their mean age (2.24 years) compared with the more recent years. These results give some credence to the suggestion that the changes observed in the mean length and mean age are real and not artefacts caused by the relatively small number of fish available to be sampled in some recent years.

5.3.9 *Rate of migration*

The rate at which smolts migrate is difficult to estimate with any degree of precision since the actual distance travelled between the points of release and recapture is not precisely known. Nevertheless, allowing for the crudeness of the experimentation, the rates at which smolts in the R. North Esk travelled in each season in 1964–69 were remarkably constant, ranging between 0.84 and 1.37 km day^{-1}. These values are consistent with the rates recorded by Mills (1989) for smolts migrating in a number of Scottish rivers.

5.3.10 *Smolt production*

Although estimates of the number of smolts 100 m^{-2} of river are recorded for

Table 5.2 Densities of salmon smolts recorded for various river systems.

River system	Density (per 100 m^2)	Authority
Canada		
Miramichi	4.7	Elson (1975)
Pollet	6.0	Elson (1975)
Big salmon	3.9	Jessop (1975)
Matamek	2.6	Gibson and Côté (1982)
Ireland		
Foyle	8.4	Elson and Tuomi (1975)
Norway		
Örkla	4.1	Garnås and Hvidsten (1985)
Vardnes	2.9	Berg (1977)
Scotland		
Bran	3.5	Mills (1964)
Shelligan	10.0−22.0	Egglishaw (1970)
Tweed	11.6	Mills *et al.* (1978)
Girnock Burn	7.0	Buck and Hay (1984)
Sweden		
Ricklea	1.9	Österdahl (1969)
USA		
Cove Brook	3.6	Meister (1962)
Wales		
Wye	4.3	Gee *et al.* (1978)

Source: Mills (1989)

one location in Ireland and four in Scotland (Table 5.2), the only complete river system in the UK for which estimates of the annual total smolt production are available over an extended period is the R. North Esk. Production varied between 93 000 in 1976 and 275 000 in 1964 (Fig. 5.4). Data are not available for all years in the period 1964−88 because high river levels during the smolt migration in some years rendered the smolt estimation technique impossible. Although fluctuating widely between years, there was no significant increase or decrease in smolt production with time (estimated correlation coefficient 'r' = −0.17, degrees of freedom = 13). No trend in the number of smolts counted in 1966−83 passing through a trap on the Girnock Burn was found by Buck & Hay (1984).

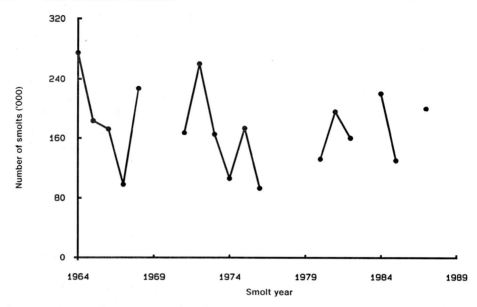

Fig. 5.4 Estimated annual smolt production in the R. North Esk, 1964–87.

5.3.11 *Predation on smolts*

Predation continues throughout the parr and smolt stages, mostly by avian predators such as various gulls (*Larus* spp) and terns (*Sterna* spp), two species of sawbill ducks (*Mergus* spp), and the heron (*Ardea cinerea*), especially when negotiating the shallow fords which are characteristic of Scottish rivers. Particular concern is being expressed at the moment at the increased distribution of the goosander (*Mergus merganser*) in Scotland and into north-west England and Wales. Anderson (1986) considered that as a direct consequence of the merganser (*M. merganser Americanus*) control programme on the Restigouche river system over a three year period, about 170 000 smolts and 8400 adults were preserved. These fish would otherwise have been lost to predation. However, Anderson found it difficult to find unequivocal correlations between presumed smolt increases and the numbers of smolts and returning adults which could be ascribed to the control of mergansers.

A simple steady-state model was developed by Shearer *et al.* (1987) to predict the effect of predation of smolts by sawbill ducks, whether in natural numbers or 'controlled', on the number of adult salmon returning to the R. North Esk. These calculations suggest that the maximum conceivable benefit from controlling sawbill duck predation was a 35% increase in the number of adult salmon returning to the R. North Esk. However, in reality, the benefit was likely to be less than this because it is improbable that all sawbills could be eliminated. Cormorants (*Phalacrocorax carbo*) feeding on the R. Bush in Northern Ireland consumed an estimated 1083–2023 wild smolts, 176–382 hatchery-reared smolts and 691−1285 brown trout each day (Kennedy & Greer 1988). Because the migration of smolts may continue for at least 28 days, this level of predation could have a big effect on the number of adults returning to the system.

Smolts which have to pass through lochs on their migration to the sea may be particularly vulnerable to predation by large trout (*Salmo trutta*) and pike (*Esox lucius*). Mills (1964) estimated that 10% of the smolt production of the R. Bran, Ross-shire, in 1959 and 1961 was eaten by pike. During the smolt migration season in the R. North Esk in some years, brown trout weighing little more than 0.5 kg have been caught with the remains of up to 20 smolts in their stomachs.

5.3.12 Smolt behaviour in estuaries

In each year between March and June, millions of smolts must leave the rivers of England, Scotland and Wales and head into the unknown. These fish have already become adapted to marine life so that salinities hyperosmotic to smolt blood present no barrier. As a result of observing fish carrying ultrasonic tags, Tytler *et al.* (1978) found that movement through tidal reaches is related to the characteristics of the estuary of each river system. In an estuary like that of the R. North Esk, where a wedge of salt water extends upstream along one bank, the pattern of smolt movement is likely to be strongly influenced by the tidal changes in the direction and velocity of water flow.

Smolts tend to leave the estuary on an ebb tide. In order to make use of the current, smolts migrate through the estuary in stages. They rest in areas of low water velocity when currents are not in the direction in which they wish to travel. However, in estuaries where the salt and fresh water are in two layers there may be a continuous downstream flow in the fresh water layer independent of the tidal cycle. In such estuaries, the pattern of movement may be continuous and similar to that found above the head of tide. Smolts are, therefore, likely to migrate through this type of estuary faster and as a result be less available to many species of predators.

5.3.13 Stock enhancement

When catches are declining, owners of fisheries, sometimes acting individually or collectively through the district salmon fishery board, have attempted to enhance the number of juveniles migrating to sea. These enhancement programmes have been carried out without any attempt to identify why catches had declined and whether there had been a comparable decline in stocks or in smolt production or in both. Such schemes have been popular at various times since the mid-1800s. Later, their popularity waned, leading to the closure of hatcheries and rearing facilities which in some cases produced smolts. In the late 1800s, for example, when catches of salmon and grilse were declining, the Spey District Salmon Fishery Board agreed that this decline could be arrested by a massive injection of reared smolts. In 1872, the Duke of Richmond and Gordon released 600 000 smolts annually from a hatchery and rearing station at Fochabers. Six years later, further rearing capacity was built at Cuningham, near Tugnet, and facilities for holding brood stock were built at Fochabers. However, net and coble catches did not improve and, in fact, declined further. In this case, releasing large number of smolts did not appear to be the solution.

A similar experiment was done on the R. Tay under the guidance of Robert Buist, Superintendent of the Protection of the Tay Salmon Fisheries. In 1853, a hatchery and smolt rearing ponds were built at Stormontfield on Scone Estate. The operations took a few years to develop and it was not until 1862 that smolts were released into the R. Tay on a regular basis. The annual production of parr or smolts was not reported except for an estimate of 120 000 smolts released into the R. Tay in 1855. The number of ova laid down appears to have varied between 60 000 and 400 000. The experiments at Stormontfield were brought to an end when it was realized that the artificial propagation of smolts conferred no real benefit on the river where natural spawning produced smolts in their tens of thousands (Thomson 1979).

Instead, the removal of obstacles in the river was begun by the proprietors themselves. Later, the district salmon fishery board, which had been constituted in the interval, took over the promotion of river improvements on the Rivers Earn, Tummel, Ericht and Isla (all tributaries). Almost 100 years were to elapse before hatcheries were again in use on the R. Tay and then only as a remedial measure for the re-seeding of spawning tributaries denuded by hydro-electric and other developments.

A somewhat similar sequence of events occurred on the R. Dee, where the original hatchery sited at Drum was abandoned. Although stocking with 'green ova' had taken place over a period of some 20 years from the late 1940s, the new hatchery at Dinnet was not built until the late 1960s at a time when the problem of ulcerative dermal necrosis was reaching its height. Appreciable numbers of fish were dying before spawning. Once again, there were no quantitative data to indicate whether or not the enhancement schemes which were undertaken had benefited the river.

From the late 1940s to 1960s, enhancement projects became fashionable once again. Hatcheries sprang up on nearly every river in Scotland. Not all the eggs were returned to the original river, either as eyed ova or unfed fry, because some district salmon fishery boards realized that the sale of eggs to other rivers and the rapidly expanding fish farming industry could bring in a substantial income — money which could be used to stabilize the assessment levied annually on each fishery proprietor. Although many thousands of unfed fry have been released annually in Scottish rivers since the 1950s, there is no documented evidence of any benefit, either in terms of increase in catch or potential spawners.

On the R. North Esk, the problem of enhancing salmon stocks after the Second World War was handled rather differently. A careful study of the physical difficulties encountered by salmon endeavouring to ascend various obstacles in the R. North Esk, both natural and man-made, was made in 1944. As a result, a programme was initiated to instal modern fish passes in both Morphie and Craigo weirs, a fish pass was cut through the rock adjacent to the Loups of the Burn (Plate 4) and some minor falls were blasted a short way upstream and on the R. West Water (Fig. 3.2). On this last fall, there is now an excellent ascent for salmon on both sides of the river at all water levels.

The falls on the R. West Water were first blasted and substantially improved in 1947. Craigo fish pass and the pass cut through the solid rock at the Loups of the Burn were completed in 1949 and the new pass at Morphie Dyke was built in 1950 with further improvement to the entrance on two occasions in the early 1980s. The total cost of the improvements prior to 1980 was under £6000. Since the completion of the main improvements, a continual endeavour has been made to improve minor obstacles on some of the spawning tributaries. Where blasting was not possible, many miles of inaccessible burns were seeded with ova and fry.

As a result of this continuing policy, in years of normal rainfall, redd counts indicated that the smaller burns near the headwaters carried a vastly increased spawning stock and fry were now being reared in many streams where none or few were present before (Smart 1965). The in-river net and coble catches also increased after these major improvements and although catches elsewhere in Scotland also increased, their overall increase was only a fraction of that of the R. North Esk fishery (Fig. 5.5). This success story, compared with the indifferent results described earlier illustrates the need to increase the availability and the amount of nursery areas and the quality of the water before attempting any enhancement. Therefore, buying off the commercial fishery without extending or improving the spawning and nursery areas can be compared with putting the cart before the horse. It is being realized that the present egg deposition by salmon in most Scottish rivers is unlikely to be limiting smolt production. However, additional egg deposition may be required if the nursery areas are to be extended.

5.4 Summary

The number of eggs produced by a female salmon depends mostly on the length and weight of the fish when it entered fresh water from the sea. The number of eggs per female salmon also varies from river to river and between different tributaries of the same river.

In 1981−89, the potential egg deposition in the R. North Esk was 14−29 million, equivalent to 6.6−13.3 eggs m^{-2} of habitat suitable for rearing fry. These densities were greater than those for most other rivers for which data are available.

Most fry disperse downstream from a redd. The distribution of redds in a tributary can be as important as the total number of redds. Maximum stocking with young salmon only occurs when the returning adults have free access to

LIVERPOOL JOHN MOORES UNIVERSITY

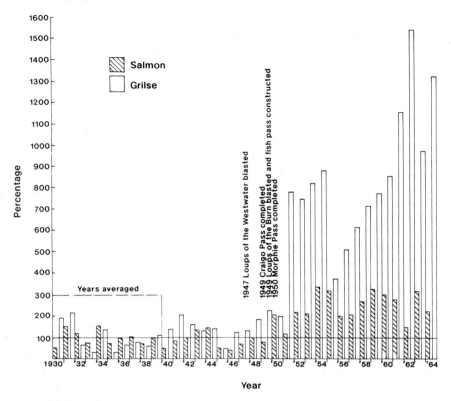

Fig. 5.5 Salmon and grilse catches by rod and by net in the R. North Esk in 1930−64, expressed as percentages of the mean 1930−39 catch. (Reproduced from Smart 1965)

the upper reaches of each tributary. However, densities of juvenile salmon differ considerably both within and between streams. The availability of food and space rather than numbers of eggs sets the main limits to fish production.

In most headwater tributaries of the R. North Esk where the quality of the rearing habitat did not change, population densities of parr of 9 cm or greater fork length did not change over 20 years and remained at 12−32 parr 100 m^{-2}. However, due mainly to bank erosion there was a reduction in the quantity of the rearing habitat in some tributaries, particularly the Water of Mark. In addition, holding pools have disappeared due to gravelling. The number of parr reared may decrease in future.

Significant differences in the mean age of the parr (9 cm or greater fork length) were found between tributaries even where parr densities were similar. These differences were reflected in the fresh water ages of the adults spawning in these tributaries and the date when they entered fresh water.

Mortality rates are highest in the weeks immediately following emergence from the gravel. Starvation is an important cause of death. The effects of predation are less clear because predators may concentrate on the fish which are shortly to die anyway from starvation and disease. Only about 1% of the eggs survive to become smolts.

Some male parr mature and stay close to the adult female grilse and salmon during redd construction. At spawning, they release their milt at the same time as the returned migrant males. In 1987, male parr fertilized 1.4−25.2% of the eggs in individual redds in the Girnock Burn. The proportion of precocious parr is less where the adult population is dominated by grilse.

In southern rivers draining into the Atlantic Ocean, Irish Sea or English Channel, most smolts go to sea at one year old. In northern Norway, smolts aged seven years are not uncommon. In Scotland, four annual age groups are normally represented in most smolt populations.

In the R. North Esk, smolts emigrate each year in April, May and June. The older age groups generally emigrate earlier than younger ones and over the whole run, female fish predominate.

In the R. North Esk, production of smolts varied between 93 000 in 1976 and 275 000 in 1964, but there was no significant trend towards increase or decrease in 1964−87. The estimated smolt production in 1989 and 1990 was 140 000 and 175 000 respectively − figures which fall within the range established since 1964.

While on passage to the sea, salmon smolts are particularly vulnerable to predation from bird predators. Where they have to pass through lochs, they may also be particularly vulnerable to predation by large trout and pike. The maximum conceivable benefit from killing all sawbill ducks on the R. North Esk would be a 35% increase in the number of adult salmon returning to that river. However, it would be impossible to kill all the ducks and it is probably not worthwhile to kill a variable smaller proportion.

Each year, millions of smolts must leave the rivers of England, Scotland and Wales. Because these fish have already become adapted to life in the sea, the increased salinities do not present a significant barrier to their migration.

In most rivers, present evidence suggests that unless there is an increase in the quality and/or size of parr-rearing habitat, any additional egg deposition or release of hatchery-reared fry or parr will be wasted and will not increase smolt production or ultimately catches.

When enhancing a river system where salmon still occur, native fish should be used as brood stock, even if reared elsewhere. All stocked fish should be marked, to monitor progress.

CHAPTER 6
LIFE IN THE SEA

Until the early 1960s, knowledge of the sea phase of the salmon's life history was mainly limited to three pieces of information. First, it was known that some smolts which migrated to sea each year returned to their home river having spent one to four winters in salt water. Secondly, the sea phase was thought to be a period of high mortality; less than 2% of the tagged smolts leaving rivers each year were subsequently recaptured as adults on return to home rivers. Thirdly, growth during the sea phase could be extremely rapid; smolts weighing on average 30 g when they left Scottish rivers in spring were able to attain a mean weight of 2.5 kg within 15 months. If these fish had remained in the sea for a further two months, they could have reached weights of between 4 and 6 kg and even perhaps 8 kg.

Where salmon went in the sea was very much a mystery. Little could be gleaned from the few salmon caught in the open sea because the site of recapture was usually a recognized fishing ground for either pelagic or demersal fish and it was known that large numbers of salmon did not feed there. Balmain & Shearer (1956) summarized the Scottish records of salmon caught at sea over the period 1880–1954. They described the capture of 90 'clean' salmon, ten salmon kelts, five grilse and three pre-grilse. Records of pre-grilse captures recall the 19 small salmon, 20–25 cm (8–10 in.) in length, captured between 28 July and 28 August in Sandrid Fjord, Ryflke, Norway (recorded by Dahl and quoted by Menzies 1914). Calderwood (1930) states that the fish described by Dahl were caught in mackerel nets and that herring (*Clupea harengus*), sand eels (*Ammodytes* spp) and sparlings (*Osmerus eperlanus*) were found in their stomachs. He also mentions the capture of pre-grilse stages off the north of Spain by fishermen using prawns as bait for the capture of bass (*Dicentrarchus labrax*). Camino (1929) refers to the capture of young salmon and sea trout in fixed nets set to catch garfish (*Belone belone*). Finally, Vibert (1952) records the capture of three salmon smolts (two in one year and one in the following year) at approximately the same position some 80 km off the south-west tip of Brittany. These smolts had been tagged in the Gave d'Oloron, south-west France, 22–27 days prior to capture at sea.

No one came nearer to describing where salmon feed in the sea than the late Mr W.J.M. Menzies who, towards the end of his 1947 Buckland lectures, stated that 'the feeding ground for the salmon of the rivers of the west of Great Britain and the north and east of Scotland is to the west or north-west in the Atlantic'. He based this conclusion on a detailed study of the movements of salmon in Canadian, Norwegian and Scottish coastal waters and the then almost complete absence of salmon smolts among the catches of other fish (e.g. herring or mackerel (*Scomber scombrus*)) except at two localities, the Bay of Fundy and a small number of Norwegian fjords.

Little did Mr Menzies imagine that the first direct evidence that the salmon on the west coast of Greenland had migrated there from elsewhere would be obtained within the next ten years. A salmon stripped at Loch na Croic on the R. Conon system in Ross-shire, in November 1955, and then tagged by the author was caught in October 1956 south of Maniitsoq, west Greenland (Menzies & Shearer 1957). Four years later, a salmon tagged as a smolt in the estuary of the Miramichi River, New Brunswick, Canada was recaptured in the same area as the Scottish tagged fish. Since 1960, salmon tagged in most of the Atlantic-salmon-producing countries have been recaptured off the west coast of Greenland.

Much more has become known about the salmon's ocean life since the high-seas fisheries started off west Greenland in 1957, and off the Faroes and in the Norwegian Sea some ten years later. This increased knowledge is due not only to the recapture in these fisheries of fish which had been tagged as smolts in home rivers but also to the recapture in home waters of adults which had been tagged in these fisheries.

6.1 Distribution at sea

6.1.1 Home waters

Various attempts have been made in recent years to catch salmon smolts in the sea. Success has been limited. Some smolts were caught by pair-trawling at night near the surface 15 km off the Kintyre Peninsula in south-west Scotland (Morgan *et al.* 1986). In addition, single smolts were caught in May in successive years by drift netting in Montrose Bay during four smolt migrating seasons in the early 1980s: 17 mm mesh multi-filament nylon nets of 5 m in depth were used. The small size of this catch, considering the large number of smolts which must leave all the rivers near Montrose and the considerable annual fishing effort, supports the views expressed by Reddin (1988) that the smolts must move rapidly away from their natal streams. For a number of years in the late 1960s and early 1970s, tags which had been attached to smolts leaving the R. North Esk were recovered from a ternery at the Sands of Forvie, near Aberdeen. Since this ternery is too far from the R. North Esk for the terns to have caught the smolts in the vicinity of the mouth of this river, the smolts must have been caught at sea somewhere between the two locations. Because no tags were returned from terneries to the south, this limited information suggests that some smolts travel north when they leave the R. North Esk and that some smolts travel relatively close to the coast and relatively near the surface. Otherwise the terns would not have been able to catch them. Apart from these few records, two single pre-grilse have been caught off

Blackpool and one off Fraserburgh. However, the whereabouts of the bulk of the post-smolts during their first summer in the sea is still uncertain.

6.1.2 The Faroese area

In the historical records describing the Faroese fish fauna, both salmon and sea trout are recorded (Svabo 1782, Landt 1800), but neither was considered to be a native species (Joensen & Vedel Taning 1970). The same authors reported that Icelandic salmon fry were released in a number of Faroese rivers in 1947−51 and salmon now spawn regularly in some of them. Although salmon were seen and caught in fresh water, the number taken annually at sea rarely exceeded five fish. However, in 1969−76, an exploratory fishery in spring and summer close to the Faroe Plateau (between 61°00′N and 63°30′N and 4°00′W and 10°00′W) succeeded in catching salmon. These were mainly 1SW (62−91%) or 2SW (7−30%) fish, but included some 3SW salmon and previous spawners (3%) (Struthers 1981). Catches in the autumn (after 31 October) and winter indicated an immigration of post-smolts into the area and there was some year-to-year variation in the relative strength of this component (Jákupsstovu 1988).

In 1969−76, 1946 salmon caught in these exploratory fishings off Faroe were tagged. Subsequently, 90 were recaptured: 33 in Scotland, 31 in Norway, 15 in Ireland, eight in other European countries and three in west Greenland. Of the fish tagged, 1650 were 1SW salmon and of the recaptures, 72 were taken in the same year in which they were tagged and seven in the following year. The dates of recapture of the remainder are unknown. The recapture of the three fish at west Greenland, two of them grilse and the other a previous spawner when tagged, established the first link between the two feeding areas.

Between 1978 and 1982, exploratory fishings took place over a much wider area. Salmon were caught from 62° to 72°N. In contrast to the earlier fishings, catch samples were now dominated (*c* 90%) by 2SW. The estimated proportion of fish in the catch native to each country also changed and fish from Scotland appeared no longer to be making the largest contribution. Based on the number of recaptures per 1000 smolts tagged in each country, the countries whose fish were now contributing most fish were Sweden, Norway, England and Scotland in that order (Table 6.1). However, as smolt production from Swedish west coast rivers is less than that of Norwegian rivers, the Norwegian contribution to the catch was greater.

Salmon tagged as smolts in Iceland, Canada, the USSR and USA have also been recaptured in the Faroese fishery. Because of the lack of additional information, the significance of the recapture of fish tagged in the USSR cannot be established. Reddin (1985) suggests that the Canadian salmon should be regarded strays. Nonetheless, the now annual occurrence of Canadian and American fish in catches may suggest that some positively migrate to the area of the Faroes. Although fish of Irish origin were as numerous as English fish in the experimental catch, few Irish fish appear in the actual commercial

Table 6.1 Summary of the total numbers of tagged salmon released by country since 1978 and the number of recoveries from the same releases in the Faroese salmon fishery.

Country	No. released	No. recaptured	No. recaptured per 1000 fish released
Sweden	60 200	302	5.02
Norway	306 500	979	3.19
England	12 200	14	1.15
Scotland	68 800	69	1.00

Source: Anon (1986a)

catches because the majority are 1SW and would be less than 60 cm long — a length below which it is a statutory offence to land fish. These fish are discarded and a proportion may survive.

A plot of the recapture sites of fish from the various countries in each statistical rectangle suggests that the salmon originating from the different countries are well mixed in the Faroese zone with no segregation by country (Anon 1984a). However, salmon stocks may not be randomly mixed within this area because in some years the number of tags recovered per 1000 fish caught tended to be higher towards the north and west than to the south of the Faroes. One possible explanation for this apparent inconsistency is a change in the annual distribution of 1SW and MSW fish due to a shift in the position of the 4°C isotherm.

The distribution of both 1SW and MSW fish in the Faroes area appears to be linked with the surface water temperature. MSW fish dominate the catches taken where the surface temperature is less than 4°C (Fig. 6.1). The availability of 1SW and MSW fish is thought to have changed both within and between years depending on the position of the 4°C isotherm (Jákupsstovu *et al.* 1985). This supports the views of Martin & Mitchell (1985) regarding the influence of temperature on the age of return of adult salmon. The mean weight of the fish caught was *c* 3.5 kg, varying slightly between years, and the catch in 1968−89 ranged between 5 and 1025 t (Table 6.2).

Analysis of the sites of recapture of fish tagged as smolts from the same group and released at the same site and on the same date suggests that fish do not, or only to a limited extent, stay together as a group/school throughout their sea life. Furthermore, cluster analyses of data from salmon caught by long-line do not indicate any significant schooling of salmon during the fishing season. Sex ratios in the sample catches taken in the Faroese salmon fisheries are biased towards females (Table 6.3). Although there is some evidence of a sex-ratio trend associated with date of capture, latitude is a much more significant factor. As the fishery moved north, the female component of the catch/1000 hooks increased significantly. These data suggest that the distribution of the two sexes within the fishery is unlikely to be

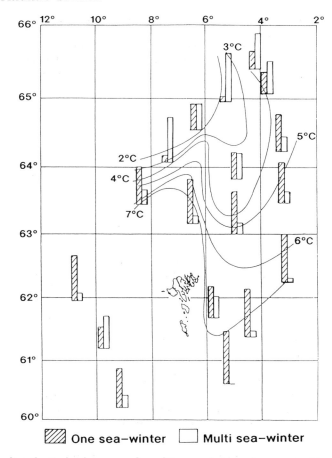

Fig. 6.1 The distribution of one- and multi sea-winter salmon caught at each fishing station in March–April 1985 together with the surface isotherms (°C). (Reproduced from Jákupsstovu *et al.* 1985)

uniform. As sex ratio is not considered when calculating quotas, potential lost egg deposition in home waters is considerably greater than that assuming equal numbers of males and females in the catches. This finding also applies to the Greenland fishery.

The number of fish of each sea age-group caught/1000 hooks in the Faroese fishery suggests that salmon probably occur in the Faroese zone throughout the year (Fig. 6.2).

Not all the fish caught in any one year would have returned to home waters in the same year. Analyses of blood serum samples for steroid hormones suggested that the proportion of fish in each sea age class which would mature in the same year would be 78% (Anon 1984a). However, in view of the difficulty of distinguishing between elevated and basal hormone levels, this value is more likely to be an underestimate than an overestimate of the true value. In addition, a greater proportion of the older rather than younger sea

Table 6.2 Nominal catches of salmon in the Faroese long-line fishery in 1968–89 (in tonnes round fresh weight).

Year	Catch	Quota
1968	5	
1969	7	
1970	12	
1971	0	
1972	9	
1973	28	
1974	20	
1975	28	
1976	40	
1977	40	
1978	37	
1979	106	
1980	553	
1981	1025	
1982	865	750
1983	678	625
1984	628	625
1985	566	625
1986	530	625
1987	576	597[1]
1988	243	597[1]
1989	364[2]	597[1]

[1] Quota is 1790 t (i.e. 597 t/year) over the 3-year period 1987–89 with an allowance for a 5% over-run on the annual average catch.
[2] Provisional figure

Factor used for converting landed catch to round fresh weight in fishery is 1.11

Source: Anon (1990a)

age classes might be expected to return in the same year.

In the 1984/85, 1986/87 and 1988/89 fishing seasons, 3–8% of the catch was estimated to be of non-wild fish which had spent a part of their previous life in a hatchery or fish farm or most probably both. In 1985/86, the range was 0–13% (Anon 1988, Anon 1990a).

Based on rather limited tagging data, the salmon present each year in the Faroese zone appear to adopt three life history strategies: some mature and leave the area for their home-rivers, some migrate as immature fish to feed near west Greenland, and the remainder probably remain to feed in the vicinity of the Faroes. We do not know whether all fish of European origin present each year near Greenland have previously passed through the Faroese zone or whether some have come directly. Whatever the route taken, fish begin arriving in the Greenland area from late summer and mix with fish from the North American (and the one Greenlandic) rivers.

Table 6.3 Sex ratio (females:males) in catches taken in the Faroese fishery during the 1981/82 season.

Date	Area fished	Sex ratio (females:males)	No. of observations
1−12.12.81	63°08′N−64°04′W 08°13′W−09°38′W	2.9:1	475
13−18.12.81	62°27′N−62°34′N 10°14′W−10°45′W	3.2:1	480
5−6.1.82	63°55′N−64°06′N 06°50′W−07°11′W	0.5:1	66
7−22.1.82	64°29′N−64°57′N 05°16′W−05°45′W	4.0:1	1051
23−31.1.82	65°50′N−66°34′N 04°28′W−05°24′W	8.0:1	742
1−4.2.82	66°16′N−66°34′N 05°17′W−05°27′W	10.1:1	274
15.2.82	67°00′N 04°00′W	58.0:0	58
16−25.2.82	67°17′N−67°51′N 02°12′W−03°23′W	108.0:1	436
24−27.2.82	68°10′N−68°11′N 01°04′W−02°40′W	64.1:1	65
1−13.3.82	70°13′N−70°45′N 02°20′W−03°18′W	418.0:1	419
20−27.3.82	64°55′N−65°12′N 01°00′W−02°00′W	2.2:1	357
17−21.4.82	66°36′N−66°43′N 02°33′W−02°53′W	64.0:1	65
24−28.4.82	64°47′N−65°12′N 06°15′W−06°34′W	35.0:0	35

6.1.3 The Greenland area

Atlantic salmon are now to be found in coastal waters around Greenland from Ammassalik (66°N) on the east coast to Kangerluk (70°N) on the west (Fig. 6.3). Both the numbers of Atlantic salmon and the localities where they can be caught off Greenland have increased since 1780 when salmon were first recorded in these seas. Around 1935, salmon occurred in fairly large numbers in the Qaqortoq, Nuuk and Sisimiut districts and by the 1950s at Nanortalik, Paamiut and Aasiaat (Shearer & Balmain 1967).

Atlantic salmon also occur in the Irminger Sea and were caught at several places there by research vessels in the summers of 1966, 1973−5, and 1985.

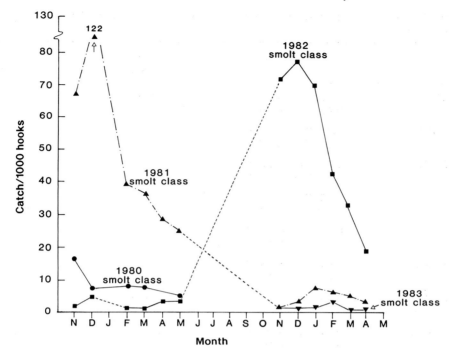

Reproduced from Anon, 1985

Fig. 6.2 Catches of salmon belonging to the 1980, 81, 82 and 83 smolt year classes taken per 1000 hooks in the Faroes fishery, November 1982 to April 1984. (Reproduced from Anon 1985)

Adult salmon tagged as smolts in North American and European rivers have been recaptured in east Greenland. An adult tagged as a smolt in the R. North Esk was caught near Ammassalik at a depth of 50 m by a fisherman fishing through a hole in the ice.

The first indication that most salmon caught off the west coast of Greenland might not be native came in 1953 from the examination of samples of scales taken from fish caught at Napasoq. Salmon caught there had experienced a much shorter river life than others caught in the R. Kapisillit. The distances involved for salmon from Europe and North America in travelling to this area would be no obstacle; tagging experiments have shown that salmon are capable of travelling distances in excess of 3200 km.

Between 1965 and 1972, 4652 salmon caught in the commercial fishery at west Greenland were tagged and released. Recoveries of tags from salmon tagged here and of others tagged as smolts in home rivers show that some salmon from nearly all the salmon-producing countries around the North Atlantic migrate to Greenland. Some of these fish, if not caught, will return to home waters.

Fig. 6.3 Map of Greenland showing places mentioned in the text.

The countries contributing to the salmon stock around Greenland are USA, Canada, Iceland, Norway, Sweden, Denmark, Finland, Scotland, England, Wales, Ireland, France and Spain. There are probably salmon in Greenland waters from the USSR and Portugal as well. The main contributors, however, are Canada and Scotland (Møller Jensen 1980).

Another interesting finding from these tagging experiments was that the contribution to the stock in Greenland waters may not be the same from all rivers producing MSW fish in a particular country. Even within the larger river systems, some tributaries may contribute more than others. For example, in the R. Tay system, the contribution made by fish originating in the R. Tummel is significantly greater than that made by fish from the R. Almond because the latter produces mainly grilse. As a result, information on the movements in the sea obtained from tagging smolts in particular tributaries of a river system may not be representative of the movements of fish from the whole of that system.

Unlike a proportion of the salmon caught in the Faroese area, all the fish caught at Greenland, apart from the few native fish, have already spent at least one winter in the sea. The majority (*c* 90%) will be 2SW salmon when caught in home water fisheries. Some salmon will overwinter in the sea, return to Greenland in the following year, and be 3 or more SW fish when they return home to spawn. It is thought unlikely that they overwinter in the Greenland area. Sex ratios in the sampled catches taken in the west Greenland salmon fishery show a bias towards females (mean *c* 3:1). Within fishing seasons, sex ratios varied between age groups, the most significant bias occurring in the 2SW component of the catch. However, in most years, this sea age group contributed less than 10% to the total annual catch. The mean weight of the salmon caught at Greenland was *c* 3.0 kg, varying slightly between years.

Various methods have been used to discover the origin of the fish caught at Greenland. Serum transferrin polymorphism was used by Payne (1980) and electrophoretic analyses of serum proteins by Child (1980), whilst the value of parasites as biological tags was investigated by Pippy (1980). A more recent method examined levels of DNA polymorphism within Atlantic salmon stocks and the use of mitochondrial DNA as a genetic marker for distinguishing North American from European stocks. Another method utilized genetic protein variation at eight loci on the DNA molecule in order to classify salmon by continent of origin. Although some salmon (primarily European) are not distinguishable directly by this method, the analysis indicated that acceptable (<6%) misclassification rates would occur using a relative likelihood model to assign specific genotype combinations to one continent or another.

Yet another method made use of the difference in shape of otoliths (the bone in the ear) from North American and European salmon. The most interesting development, however, was the application of the discriminant function analysis of scale characters (Lear & Sandeman 1980). Further refinement of the method allows also for a separation of hatchery-reared and wild fish (Anon 1986a). In 1969–89, the estimated proportion of European salmon present in the catch at west Greenland varied between 38% and 66% (Table 6.4).

In 1983 and 1984, the landings of salmon at Greenland (310 and 297 t) were the lowest recorded since 1962 (Table 6.5). This was a relatively greater decline than any experienced by other fisheries in the North Atlantic during these two years. Three factors have been suggested as the possible cause: adverse environmental conditions; low abundance of juveniles, coupled with a below average survival of the relevant smolt cohorts, and a reduction in the fishing effort at west Greenland. In 1983, cod also practically disappeared from the fishing banks at west Greenland and this sudden change could not be

Table 6.4 Percentage (by number) of North American (NA) and European (E) salmon in research vessel catches at west Greenland (1969−82) and from commercial samples (1978−89).

Source	Year	Sample size		Continent of origin (%)			
		Length	Scales	NA	(95% CI)[1]	E	(95% CI)
Research	1969	212	212	51	(57,44)	49	(56,43)
	1970	127	127	35	(43,26)	65	(74,57)
	1971	247	247	34	(40,28)	66	(72,50)
	1972	3488	3488	36	(37,34)	64	(66,63)
	1973	102	102	49	(59,39)	51	(61,41)
	1974	834	834	43	(46,39)	57	(61,54)
	1975	528	528	44	(48,40)	56	(60,52)
	1976	420	420	43	(48,38)	57	(62,52)
	1977	−	−	−	(−)	−	(−)
	1978[2]	606	606	38	(41,34)	62	(66,59)
	1978[3]	49	49	55	(69,41)	45	(59,31)
	1979	328	328	47	(52,41)	53	(59,48)
	1980	617	617	58	(62,54)	42	(46,38)
	1981	−	−	−	(−)	−	(−)
	1982	443	443	47	(52,43)	53	(58,48)
Commercial	1978	392	392	52	(57,47)	48	(53,43)
	1979	1653	1653	50	(52,48)	50	(52,48)
	1980	978	978	48	(51,45)	52	(55,49)
	1981	4570	1930	59	(61,58)	41	(42,39)
	1982	1949	414	62	(64,60)	38	(40,36)
	1983	4896	1815	40	(41,38)	60	(62,59)
	1984	7282	2720	50	(53,47)	50	(53,47)
	1985	13272	2917	50	(53,46)	50	(54,47)
	1986	20394	3509	57	(66,48)	43	(52,34)
	1987	13425	2960	59	(63,54)	41	(46,37)
	1988	11047	2562	43	(49,38)	57	(62,51)
	1989	9366	2227	56	(60,52)	44	(48,40)

[1] CI − confidence interval calculated by method of Pella and Robertson (1979) for 1984−86 and by binomial distribution for the others.
[2] During fishery
[3] Research samples after fishery closed.

No data are available for the years 1977 and 1981

Source: Anon (1990a)

explained on the basis of overfishing or recruitment failure. However, the two fish species lived in the same environment, and this led to the suggestion that these failures were caused by adverse environmental conditions.

The winters of 1983 and 1984 at west Greenland were extremely cold because the Arctic Canadian cold air mass moved to the Davis Strait area with its centre near Aasiaat at west Greenland. Here, temperatures in the 1983 and 1984 winters fell to 12°C and 14°C below their respective means (Rosenørn *et al.* 1985). This suggestion fits with the observations of various authors

Table 6.5 Nominal catches of salmon at west Greenland, 1960−89 (in tonnes round fresh weight).

Year	Origin of foreign fishermen				Gill net and drift net Greenland[2]	Total	Total allowable catch (TAC)
	Norway	Faroes	Sweden	Denmark			
1960	0	0	0	0	60	60	
1961	0	0	0	0	127	127	
1962	0	0	0	0	244	244	
1963	0	0	0	0	466	466	
1964	0	0	0	0	1539	1539	
1965	NA[1]	36	0	0	825	861	
1966	32	87	0	0	1251	1370	
1967	78	155	0	85	1283	1601	
1968	138	134	4	272	579	1127	
1969	250	215	30	355	1360(385)	2210	
1970	270	259	8	358	1244	2146[3]	
1971	340	255	0	645	1449	2689	
1972	158	144	0	401	1410	2113	
1973	200	171	0	385	1585	2341	
1974	140	110	0	505	1162	1917	
1975	217	260	0	382	1171	2030	
1976	0	0	0	0	1175	1175	1190
1977	0	0	0	0	1420	1420	1190
1978	0	0	0	0	984	984	1190
1979	0	0	0	0	1395	1395	1190
1980	0	0	0	0	1194	1194	1190
1981	0	0	0	0	1264	1264	1265[4]
1982	0	0	0	0	1077	1077	1253[4]
1983	0	0	0	0	310	310	1190
1984	0	0	0	0	297	297	870
1985	0	0	0	0	864	864	852
1986	0	0	0	0	960	960	909
1987	0	0	0	0	966	966	935
1988	0	0	0	0	893	893	—
1989[5]	0	0	0	0	337[5]	337[5]	900

[1] Figures not available, but catch is known to be less than the Faroese catch.
[2] For Greenlandic vessels: up to 1968 set gill net only; after 1968, set gill net and drift net. All non-Greenlandic catches from 1969−84 were taken with drift nets. The figures in brackets for the 1969 catch are an estimate of the minimum drift net catch.
[3] Including 7 t caught on long-line by one of two Greenland vessels in the Labrador Sea early in 1970.
[4] Quota corresponding to specific opening dates of the fishery.
[5] Provisional figure

Factor used for converting landed catch to round fresh weight in fishery by Greenland vessels is 1.11. Factor for Norwegian, Danish and Faroese drift net vessels is 1.10.

Adapted from: Anon (1990a)

including Cushing (1983) and Dunbar (1981) who have shown that climatic changes can affect the distribution and abundance of many species of marine fish, including salmon at west Greenland (Dunbar & Thomson 1979). Reddin

& Shearer (1987) also showed that the abundance of salmon at west Greenland was affected by the environmental conditions in January and August in the north-west Atlantic. The same phenomenon occurred in 1989: a cold winter resulting in low water surface temperatures, followed by a low salmon landing (337 t). Cold water (less than 3−4°C) can stop salmon migrating because it acts as a barrier through which they will not penetrate. In colder years, fewer salmon may move into the west Greenland area from the Labrador sea than in warmer years and fewer salmon may also migrate from the Faroese area, some preferring to return to home waters as grilse. Martin & Mitchell (1985) found that an increase in the temperature in the subarctic (Grimsey Island) was associated with larger numbers of MSW fish and fewer 1SW fish returning to the R. Dee in Aberdeenshire. This coincides with recent events in Canada where ratios of grilse to salmon have decreased coincidental with a colder marine climate (Reddin 1988).

6.2 Feeding

The first published information on the food and feeding of Atlantic salmon in the open North Atlantic was provided by Templeman (1967). On the basis of ten stomachs examined in the Labrador Sea during July and August 1965, he thought that salmon fed mainly on Arctic squid (*Gonatus frabricii*) and on *Notolepis rissoi kroyeri*. However, on the west Greenland banks, salmon fed on capelin (*Mallotus villosus*) and sand eels (Lear 1980).

Hansen (1965) reported that, when close to the Greenland coast in autumn, salmon ate mainly euphausiids and capelin with some sand eels. In 1968−70, salmon caught off the west coast of Greenland between Kangerluk and Paamiut were feeding mainly on capelin (59% of total food by weight), sand eels (14%), amphipods (9%), euphausiids (6%), fish remains (6%), *Paralepis coregonoides borealis* (2%) and other items (4%). The composition was similar for all length groups from 40 to 99 cm and the amount of food in the stomachs averaged 0.7% of total body weight (Lear 1980). Shearer & Balmain (1967) also found capelin to be the most frequent (86%) food item in 543 salmon stomachs examined in 1965 and 1966 from fish caught in three west Greenland fjords − Uqpik, Praeste and Kigdlut Iluat, south of Nuuk.

Four salmon caught by the Danish research vessel 'Dana' during drift netting experiments in the Irminger Sea at the end of June 1966 had been feeding on squid (*Brachioteuthis riisei*) and amphipods (*Themisto gaudichaudi*) (Jensen 1967).

Nearer to home in the north-east Atlantic, Pyefinch (1952) recorded that the stomach of a pre-grilse, measuring 41.5 cm, caught 15 km south-east of Fuglø Head (Faroes) in January 1952 contained amphipods (*Themisto gracilipes*), euphausiids (*Thysanvessa longicandata*) and sand eels (*Ammodytes lancea marinus*). Struthers (1970, 1971) examined the stomach contents of 132 salmon caught by long-line baited with sprats (*Sprattus sprattus*) north of the Faroes (between 62°30′N and 63°30′N) in April 1970 and a further 140 caught in

March/April 1971. Of the stomachs, 36% contained no food, 44% contained crustacea (principally amphipods of the genus *Parathemisto*), and 33% fish remains. In many cases, however, the fish in the stomachs were sprats with which the long-lines had been baited. If the stomachs containing sprats are ignored, fish were present in only about 10%. The results suggested that crustaceans constituted the main food of salmon at the Faroes.

Hislop and Youngson (1984) examined the contents of 48 stomachs taken from fish caught in March 1983 between 64°30′N and 65°30′N. The proportions of fish and crustacean prey occurring were identical (92%). Fish made up the greatest part of the contents of 30 stomachs and the principal identifiable fish prey were lantern fishes (*Myctophidae*). These were found in 40 (83%) of the stomachs. Up to 28 individuals were recorded in a single stomach. Pearlsides (*Maurolicus muelleri*) were found in 12.5% of the stomachs but these rather small fish did not make a major contribution by weight. Capelin (4 stomachs), squid (*Todarodes* spp) (2), and one barracudina (*Paralepis* spp) were also identified. In 16 stomachs, crustaceans were the main food, mostly amphipods of the genus *Parathemisto*. Frequently more than 500 individuals were found in a single stomach. Euphausiids occurred in 44% of stomachs, but their biomass was not important. Food organisms identified by Walkington *et al.* (unpublished data) in 233 salmon caught near Faroe were similar to those recorded earlier. *Paralepis* spp occurred in the stomachs of the fish caught furthest to the north and in the greatest depth of water. Sand eels and Norway pout (*Trisopterus esmarkii*) were found in the stomachs of fish caught at the shallower sites. This varied menu suggests that adult salmon are opportunistic feeders and prey on whatever organisms are present.

A total of 1145 stomachs from Atlantic salmon caught off northern Norway in late winter and spring in 1969−72 was examined by Hansen & Pethon (1985). The most important food items found in fish caught over the shelf area were euphausiids and hyperid amphipods while myctophids (*Benthosema glaciale*), squids (*Gonatus fabricii*) and euphausiids were found most frequently in salmon caught elsewhere. Most salmon had preyed on only one species. However, these stomachs represent only one day's feeding and the presence of the arctic shrimp *Hymenodora glacialis* shows that salmon can feed to a depth of at least 300 m. This might explain why Lea hydrostatic tags are sometimes flattened on recaptured salmon.

In 1895, Archer, the Inspector of Salmon Fisheries for Scotland, reported that 13% of 1422 salmon caught near the mouth of the R. Tweed in March (35%), April (24%), July (1%) and August (3%) contained food which was mainly of marine origin. Fish remains, largely herring, were most common, together with crustacea and annelids. Of the stomachs of 88 salmon caught by drift net in the North Sea in spring and summer 1964, 48% (spring) and 22% (summer) contained food showing a similar correlation between presence or absence of food and season. Herring, sprats, sand eels, polychaete worms (*Nereis* spp) amphipods and the euphausiid *Meganyctiphanes norvegica* were found in the stomachs irrespective of where the fish were caught. However, invertebrates were absent in summer and autumn samples. Fraser (1987) referred to sand eels *Ammodytes marinus* in the stomachs of some salmon caught in bag nets set off the north-west coast of Scotland. Using rather limited data, he suggested that either there is a local feeding stock of salmon in this area or that fish feed during their migration from distant waters.

Browne *et al.* (1983) examined the probable vulnerability of species such as sprat and herring on the basis that the optimal prey size in the ocean is 2.2–2.6% of fish body size. They concluded that species such as sprat apparently grow fast enough to be at risk for a relatively brief period whereas 0+ herring which grow at a much slower rate, are an almost ideal size for salmon throughout their first 12 months at sea. Morgan *et al.* (1986) found 2–6 cm sand eels in the stomachs of post-smolts (16.7–20.0 cm long) caught off the Kintyre Peninsula in June while pair-trawling at night in the surface 10 m.

The presence in significant quantities of marine species including sand eels, herring and capelin in the diet of salmon has concerned people interested in the effects on salmon of industrial fisheries. A significant reduction in the number of these prey species could conceivably affect the survival and growth of salmon (Mills 1987). However, Reddin & Carscadden (1982), who studied the possible interactions between salmon and capelin, could find no significant relationship between the abundance of salmon and capelin on which the salmon might have been feeding. They concluded that the poor survival of the 1977 smolt class from Canadian rivers could not be attributed to a recent decline in capelin.

6.3 Migration routes

Too few fish have been captured at intermediate locations to encourage speculation about the oceanic routes followed by the migrating fish, if indeed they do follow well defined routes to and from distinct feeding grounds. Royce *et al.* (1968), Stasko *et al.* (1973) and Hawkins *et al.* (1979a) all concluded that the movements of the returning fish cannot be explained in terms of passive drift with water currents. This is because the residual velocities of oceanic currents are insufficient to account for some observed speeds of migration. In the absence of such a passive mechanism, Stasko *et al.* (1973) concluded that the movements of Atlantic salmon are directed. They must be oriented to particular stimuli. A great variety of environmental factors have been proposed by different authors as providers of the necessary cues. An extreme view (Baker 1978) proposed that adult salmon return through an area familiar through previous exploration, a technique referred to by Harden Jones (1968) as pilotage. Smith *et al.* (1981) concluded that fish swimming offshore maintained a relatively constant heading and swimming speed, independent of the speed and direction of local tidal currents. An opposing view, that fish reach their destinations largely by random wandering and passive drift, was proposed by Huntsman (1934) and by Saila & Shappy (1963).

An economical mechanism by which a directional movement may take place is provided if the behaviour of the fish is different at different phases of the tide. This has been termed selective tidal stream transport (Greer-Walker *et al.* 1978). By varying the depth at which the fish swims (and hence varying the degree of passive carriage by stratified tidal currents), or by varying the speed and direction of motion relative to the tidal stream at different phases of the tide, a plaice (*Pleuronectes platessa*) may move in a particular tidal direction with a minimum of effort. There is no evidence of this for salmon.

Principally due to the lack of any systematic tagging of salmon on the high seas in the north-east Atlantic area, the migration routes of European salmon to and from their feeding grounds are virtually unknown. This is in marked contrast to the wealth of information recently summarized by Reddin (1988) describing the migration routes of North American grilse and 2SW (or older) salmon to and from their feeding areas (Figs 6.4 & 6.5). Notwithstanding the increased knowledge of marine migration routes on the west side of the Atlantic, how salmon find their way in the ocean and eventually back to the vicinity of their parent river remains a mystery on both sides of the Atlantic. Various mechanisms have been suspected as the means whereby salmon regulate their oceanic movements including an ability to sense, for example, temperature, salinity and water velocity gradients (Westerberg 1982, 1984) and the gradient in electrical potential generated by the movement of an ocean current in the earth's magnetic field (Rommel & McLeave 1973, Stasko *et al.* 1973).

May (1973) suggested that in the north-west Atlantic salmon occur in relatively cool water at 3°C to 8°C. More recently, Martin *et al.* (1984), Scarnecchia (1984), Martin & Mitchell (1985) and Reddin & Shearer (1987) all demonstrated direct relationships between ocean climate, sea-surface temperature, and the abundance and distribution of salmon at sea. Reddin & Shearer (1987) also suggested that salmon can modify their migratory path depending on the temperature of the water through which they must swim and that older salmon will tolerate colder sea-surface temperatures than younger fish. Stewart (1978) suggested that when a smolt leaves coastal waters it will come within the influence of an oceanic gyre in which it will drift along. At certain times it may leave, rejoin it or even transfer to associated gyres. The ultimate result, if it remained within its original gyre, would be to return to the point at which it embarked as a smolt. Certainly there is no lack of current systems in the North Atlantic Ocean (Fig. 6.6). But all this complex speculation does not conceal the fact that we do not know how salmon orient their movements in the sea. Perhaps this is just as well, because as soon as we do know, there will be those who will want to capitalize on the knowledge, exploit the fish and add increased mortality rates to the already complex story.

6.4 Predation

Most predation at sea is by fish predators listed by Wheeler *et al.* (1974). Sharks, in particular Greenland shark (*Somniosus microcephalus*) and porbeagle (*Lamna nasus*) together with skate (*Raja batis*), cod (*Gadus morhua*), ling (*Molva molva*), whiting (*Merlangius merlangus*), pollack (*Pollachius*

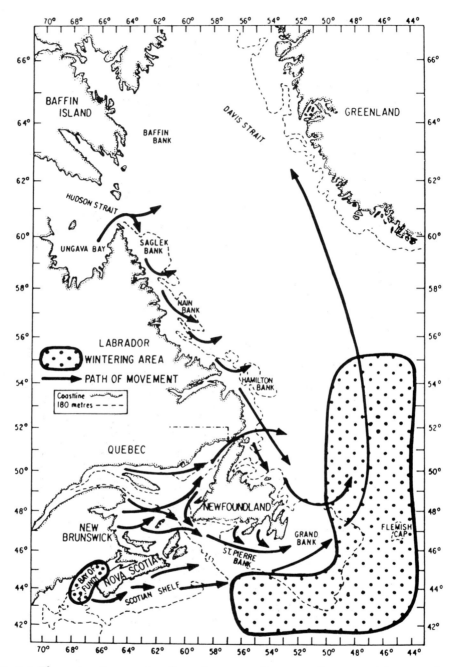

Fig. 6.4 The migration routes for salmon smolts away from coastal areas show-ing possible overwintering areas and movement of multi sea-winter fish into west Greenland. Arrows indicate the path of movement of the salmon, and shaded area indicates where the fish overwinter. (Reproduced from Reddin 1988)

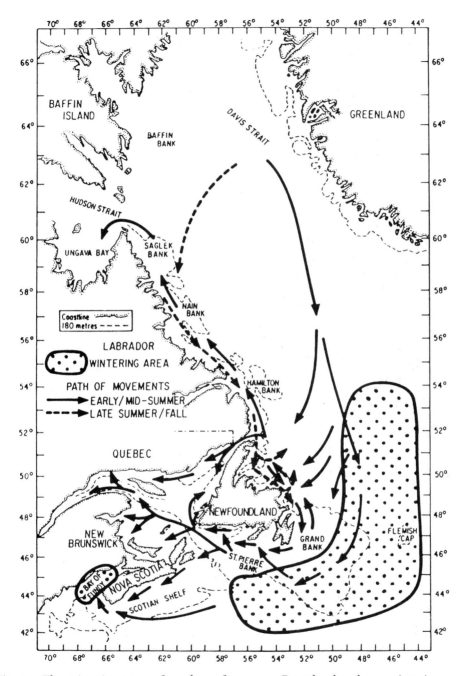

Fig. 6.5 The migration routes for salmon from west Greenland and overwintering areas on return routes to rivers in North America. Solid arrows indicate migration in mid-summer and earlier: broken arrows indicate movement in late summer and autumn: shaded areas indicate where the fish overwinter. (Reproduced from Reddin 1988)

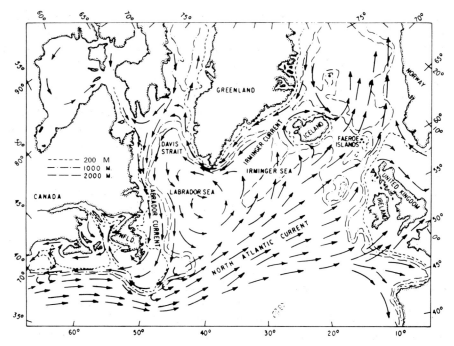

Fig. 6.6 The main surface currents in the northern part of the North Atlantic. (Reproduced from Reddin 1988)

pollachius), saithe (*P. virens*) and halibut (*Hippoglossus hippoglossus*) are known to predate on salmon. Bottlenose dolphins (*Tursiops truncatus*) have been observed tossing salmon in the air off the mouth of the R. Spey but it is not known whether the fish were actually eaten. Twenty three salmon were found in the stomachs of five sharks caught off the Aberdeenshire and Kincardineshire coasts between 1911 and 1933 (Balmain & Shearer 1956). Piggins (1959) and Hvidsten & Møkkelgjerd (1987) described how, in a restricted area, pollack and cod prey heavily on emigrating smolts, accounting for 25% of the mortality. Similarly, saithe and sea trout in Norwegian fjords and bass (*Dicentrarchus labrax*) in Northern Ireland have been recorded taking smolts (e.g. Kennedy 1954). In contrast, an examination of nearly 1000 codling taken close to the mouth of the R. Dee on the Scottish east coast during the smolt run showed no trace of young salmon (Rae 1964). However, a cod caught 5 km off Troup Head (Moray Firth) on 14 May 1964 was found to contain a 3-year-old 15 cm smolt. Rae (1966) gave two further examples, a salmon 35−40 cm long found in the stomach of a large cod caught 13 km ESE of Aberdeen in February 1964, and a 45 cm salmon in a large cod caught off Eyemouth in May 1965. The conclusion is that while cod may be potential predators on salmon smolts in Scottish waters, they do not take appreciable numbers, and the only serious attempts to assess the extent of this predation have shown it to be almost negligible. Calderwood (1907) wrote: 'The stomachs of coalfish netted in the mouth of the R. Spey have been found to be full of freshly swallowed smolts especially during May and June'. Morgan *et al.* (1986) examined the stomachs of 763 dog fish (*Squalus acanthias*) caught off the Kintyre Peninsula in June, soon after the emigration of smolts from the neighbouring rivers, and found that none had eaten smolts.

There is extremely little information on predation in the sea by avian predators. Caspian terns (*Hydroprogne tschegrava*) and various species of gull were the main predators in the Baltic (Valle 1985). Nearer home, avian predators known to take smolts in the sea include cormorants, shags (*P. aristoteles*) and various species of terns and gulls.

Predation at the post-smolt stage may be important, particularly as the number of predatory species able to eat smolts dramatically decreases as salmon grow. Above a weight of approximately 250 g, salmon are no longer available to most bird species.

There are about 97 000 grey seals (*Halichoerus grypus*) and 22 000 common seals (*Phoca vitulina*) around the UK coastline (Sea Mammal Research Unit, pers comm). Both species eat salmon. Salmon fishermen often justifiably complain about the effects of seals on their catches. There are four problems: attacks by seals on salmon at sea which can result in the complete loss of fish or damage to them; attacks on salmon caught in nets which again can result in the complete loss of fish or damage to them; loss of catch due to the presence of seals in the vicinity or actually in the net; and physical damage to the nets which can result in lost fishing time and/or complete or partial loss of catch. The problem of net damage has been much reduced since the introduction in the 1960s of nets made from synthetic fibre.

Seals do not catch and consume every salmon which they attack. Salmon bearing wounds which appear to have been inflicted by seals are regularly found in the commercial catch. Damage is of two types. 'Fresh damage' has presumably been inflicted immediately before the fish was caught or during the time it was in the net. 'Healed damage' may have been inflicted months or even years before the fish was caught. Damage is most severe in the spring at the time when salmon normally command the highest price (Table 6.6).

Shearer & Balmain (1967) recorded 2.8% of the west Greenland catch in 1965 and 1966 with either 'fresh' or 'healed' wounds consistent with attack by seals. A salmon in the Scottish catch with a 'healed' wound could have been mutilated off the Greenland coast. It is often suggested that seals require the physical presence of nets in order to inflict damage. However, the presence of freshly mutilated salmon in Kinnaber Mill trap several months after all fishing gear had been removed from the sea suggests that seals can damage salmon in the absence of all salmon nets.

Rae (1960) described the distribution of both grey and common seals and their association with Scottish fisheries. On the basis of the number of seals seen by fishermen in the vicinity of their nets and only when they were fishing, Rae & Shearer (1965) estimated that about 148 000 salmonids had been killed by seals on the Scottish coast between 1959 and 1963. Parrish & Shearer (1977) estimated that 195 000 t of fish were consumed annually by

Table 6.6 Mean monthly price of salmon (one and multi sea-winter) at Montrose in 1985–90 (£ kg^{-1}).

Month	One sea-winter salmon						Multi sea-winter salmon					
	1985	1986	1987	1988	1989	1990	1985	1986	1987	1988	1989	1990
February							8.75	9.47	7.48	10.57	10.26	12.72
March							8.00	7.06	7.81	8.26	10.50	11.35
April			5.51	5.30	5.18	4.63	7.60	5.75	7.37	7.12	8.17	8.11
May	4.57	5.52	5.56	4.49	4.02	4.61	7.00	5.02	7.37	6.49	6.14	5.76
June	4.79	4.30	5.05	4.55	3.91	4.36	6.43	4.57	6.62	5.62	5.60	5.27
July	4.45	3.65	5.01		3.45	4.15	6.12	4.09	6.29	5.47	4.88	5.07
August	5.09	4.28	5.90	5.39	4.14	5.12	5.73	4.12	6.69	5.78	4.65	5.49

grey and common seals in Scottish waters of which about 130 000 t were commercially exploited species. The total catch of all species taken within the UK's extended fisheries limits varies between 650 000 and 1 300 000 t. Grey seals eat mostly sand eels (61% by weight) followed by cod (19%), saithe (6%), unidentified flat fish (3%), haddock (3%), whiting (2%) and flounder (2%) (Anon 1984b). All other species contributed less than 1% to the diet. The total annual consumption of fish by grey seals is about 57 000 t, calculated by assuming that invertebrates and any fish whose hard parts do not find their way into the faeces make an insignificant contribution to the seals' energy requirements.

These data provide an estimate of the maximum total amount of these fish species consumed by grey seals. The data however, underestimate the contribution of larger fish, particularly salmon, of which seals tend to eat only the soft parts leaving no trace in the faeces (Plate 16 in the colour section). In addition, the contribution made by sand eels, for example, may be overestimated because an unknown percentage of sand eel otoliths in the faeces (on which the estimates of consumption are based) may have come from the guts of other fish consumed by the seals. The estimate of 57 000 t is considerably less than the estimate of Parrish & Shearer (1977), even though the seal population has significantly increased in the interval.

6.5 Natural mortality

As assessment of the natural mortality of salmon in the sea is crucial for management programmes which include the impact of high seas fisheries on home water stocks and catches. Doubleday *et al.* (1979) calculated the marine natural mortality of Atlantic salmon from the Sandhill River, Labrador and the R. Bush, Northern Ireland. Shearer (1984b) describes three models for estimating the natural mortality at sea for salmon from the R. North Esk based on data corresponding to the smolt years 1964−68, 1971, 1976 and 1980−81. In an earlier paper, Pratten & Shearer (1981) estimated the fishing mortality on North Esk salmon. A number of the equations formulated earlier by Shearer was revised, using values for the close season from the automatic fish counter at Logie.

In their investigation of fishing mortality on salmon from the R. North Esk, Pratten & Shearer (1981) first estimated the size of the North Esk salmon population returning to home waters, including the R. North Esk. In this analysis, the number of salmon reaching the furthest upstream major net fishery in the R. North Esk was estimated using a mark-recapture experiment. Known numbers of adult salmon from Kinnaber Mill trap were tagged before release into the river downstream of this major fishery.

Using data from the recapture of fish which had been tagged as smolts, the contribution of salmon from the R. North Esk to other fisheries was estimated from the ratio of marked to unmarked salmon in the North Esk net and coble catch, assuming that this ratio remained the same in all fisheries which exploit North Esk salmon. The method assumes that the rates of tag loss and nonreported tag recaptures are negligible or can be measured (this is constantly being monitored), that the proportion of the returning North Esk salmon which enters the river during the annual close time (E) is 0.10 or 0.20, and that the North Esk net and coble fishery exploits only fish of North Esk origin.

The numbers of smolts emigrating from the R. North Esk each year were estimated by a mark−recapture technique. Known numbers of smolts were tagged from Kinnaber Lade trap in 1964−68 and from Kimber Mill trap in 1971−85 and released into the main river upstream of both traps. Subsequent captures at the relevant trap, including recaptures, were used to estimate the number of smolts migrating downstream by means of a stratified population model (Schaefer 1951). This model allows the number of smolts migrating over short periods (five days) throughout the duration of the run to be estimated. Smolts and returning adults were grouped according to their total age, broken down into river and sea ages where appropriate.

Between 14% and 53% of each cohort of smolts which emigrated to sea in 1964−8, 1971−6, 1980−2 and 1984−5 returned as adults to home waters (Table 6.7). Missing data are due to the lack of reliable estimates of smolt numbers.

Table 6.7 Estimated annual smolt production and numbers of salmon returning to home waters from the smolt years 1964−8, 71−6, 80−82, and 84−5.

Smolt year	Smolt production	Year of return migration	Estimated number returning to homewaters	Percentage survival
1964	275 000	1965	37 232	
		1966	27 304	24.9
		1967	3 873	
1965	183 000	1966	28 589	
		1967	27 819	33.7
		1968	5 341	
1966	172 000	1967	44 813	
		1968	37 389	49.5
		1969	2 916	
1967	98 000	1968	34 630	
		1969	16 223	52.9
		1970	969	
1968	227 000	1969	63 384	
		1970	11 818	33.7
		1971	1 190	
1971	167 000	1972	38 291	
		1973	14 211	31.6
		1974	317	
1972	260 000	1973	63 409	
		1974	12 227	29.3
		1975	502	

Table 6.7 (Continued)

Smolt year	Smolt production	Year of return migration	Estimated number returning to homewaters	Percentage survival
1973	165 000	1974	34 932	
		1975	20 178	33.7
		1976	425	
1974	106 000	1975	36 297	
		1976	11 004	45.0
		1977	379	
1975	173 000	1976	20 400	
		1977	16 871	21.7
		1978	337	
1976	93 000	1977	25 413	
		1978	17 453	46.6
		1979	459	
1980	132 000	1981	13 695	
		1982	9 610	18.0
		1983	453	
1981	195 000	1982	27 163	
		1983	12 780	20.7
		1984	493	
1982	160 000	1983	16 190	
		1984	8 359	16.0
		1985	1 109	
1984	225 000	1985	22 158	
		1986	9 677	14.3
		1987	267	
1985	130 000	1986	31 246	
		1987	7 195	29.7
		1988	208	

Estimates of the survival rates of different cohorts of smolts showed great variation. The first four values in the 1980s indicate a marked decrease in survival compared with the mean value for the previous 11 years (t = 3.62, df = 13, P < 0.01). However, the 1985 value showed an increased survival and falls within the range established since 1964 (Table 6.7). The preliminary estimate for the 1987 smolt cohort is 20% which again falls within the established range but towards its lower end. However, the survival rate of the smolts which emigrated in 1989 and 1990 may be less than the lowest figure previously recorded in the time series. The model used to determine the estimates is described in Appendix A.

In 1963–1988, there was no significant trend either upwards or downwards in the length or weight of returning 1SW, 2SW, and 3SW fish (Figs 6.7 and 6.8). But the mean weight of 2SW fish has increased significantly. This may

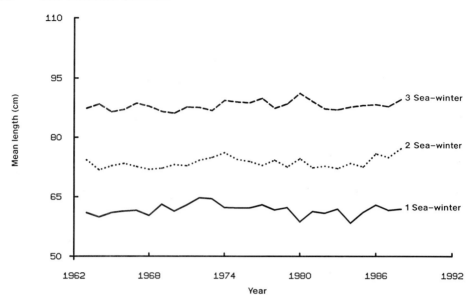

Fig. 6.7 Mean length (cm) of one, two, and three sea-winter fish returning to the R. North Esk in 1963−88.

result from a higher proportion of the 2SW catch now being taken later in the season than previously. None of the other correlation coefficients was significantly different from zero.

Observations suggest that recent changes in the numbers of adult salmon returning to the R. North Esk result from effects occurring during the sea phase. They were not the result of any changes which could have occurred before the smolts emigrated. For example, parr densities remained stable and smolt production and mean smolt age did not change significantly.

The analysis of survival from egg to smolt and from smolt to adult for salmon stocks in Western Arm Brook, Newfoundland, and spawner-recruit ratios for the Ellidaar, Iceland indicated that survival while salmon are in the marine environment was four times more variable than survival in fresh water (Reddin 1988). This also suggests that much of the fluctuation in salmon catches is related to sea survival.

Avian predators, including cormorants, shags and various species of terns and gulls, are known to take smolts emigrating from the R. North Esk but no estimate of the level of predation on smolts from the R. North Esk by these predators has been made. However, Rae (1969) concluded that cormorants and shags were unlikely to be serious predators on salmon smolts in Montrose Bay. Hvidsten & Møkkelgjerd (1987) in Norway reported that the main mortality on hatchery-reared smolts was predation by cod. However, wild smolts might not be as susceptible to predation and cod stocks have declined. Other marine fish species are known to prey on salmon but their levels of predation are not known.

If the decrease in marine survival had resulted from food shortage, it could

Fig. 6.8 Mean weight (kg) of one, two, and three sea-winter fish returning to the R. North Esk in 1963−88.

reasonably have been expected that the mean length and weight of all sea age groups returning in the 1980s would have been significantly different from the corresponding values obtained in the previous two decades. No such difference has occurred. However, this result does not exclude the possibility that the cause of the increased marine mortality is a biological factor associated with the availability of food. For instance, it is possible that an environmental factor, e.g. temperature, could have advanced or delayed the sand eel spawning season in those years of high mortality. Consequently, the 0+ sand eels would be either too big a prey or not available to the post-smolts when required. As a result, a marine mortality of post-smolts could have occurred without any change in the growth rate of those that survived.

In addition to natural mortality, including predation, the estimated marine mortality includes losses to the high seas fisheries and illegal fisheries. Unfortunately, we cannot apportion the total loss between each of these causes. Nevertheless, because catches taken in the Greenland and Faroese fisheries were smaller in the 1980s than in the 1960s and 70s, we may assume that fewer North Esk fish were caught in these fisheries in the more recent times.

From all this, it seems that the marked increase in mortality at sea observed in the 1980s, apart from that on the 1985 smolt cohort, could have been due to an increase in illegal fishing and/or predation. No reliable estimates of losses to illegal fisheries are available, and we cannot tell whether there has been an increase in non-catch fishing mortality (e.g. from netting or from seals) on salmon native to the R. North Esk in the 1980s as compared with the 1960s and 1970s.

6.6 Summary

Before the early 1960s, we knew little of the sea phase of the salmon's life history. Although this phase was known to be one of rapid growth and assumed to be one of high mortality, few salmon or smolts had been caught in the open sea or in marine coastal waters outside home waters.

The migration routes of European salmon to and from their feeding grounds, remain virtually unknown which is in marked contrast to the wealth of knowledge describing the pathways followed by North American salmon to and from their feeding areas. However, this increased knowledge has not helped to elucidate the mechanisms whereby salmon regulate their oceanic movements.

Direct relationships have been demonstrated between ocean climate, sea surface temperature and the abundance and distribution of salmon at sea. Salmon may therefore be able to modify their migratory path depending on the temperature of the water through which they must swim. Older salmon can tolerate colder sea-surface temperatures than younger fish.

The data describing the feeding habits of salmon in the sea are conflicting. Some authors suggest that adult salmon are opportunistic feeders and prey on whatever organisms are present while other studies showed a high preference for one particular type of food.

Juvenile salmon have been found in the stomachs of many marine species, and predation at the post-smolt stage may be important, particularly as the number of predatory species able to eat salmon dramatically decreases as the salmon grow. Seals eat salmon. Because seals do not swallow the bones, the contribution which salmon have been found to make to the diet of seals is a major underestimate. The marine survival of different cohorts of smolts emigrating from the R. North Esk in 1964–85 varied between 14 and 53%.

CHAPTER 7

THE HOME WATER FISHERIES

7.1 Ownership in Scotland

All rights of salmon fishing in Scotland, whether in the sea, in estuaries or in rivers are *inter regalia*, and originally belonged to the Crown. However, in many cases, these rights have been conveyed by means of written Crown grants to individuals. All private titles to salmon fishing are derived from such grants and must be based on deeds recorded in the Register of Sasines, kept at Meadowbank House, Edinburgh. In certain cases, the unchallenged possession of salmon fisheries for the appropriate prescriptive period may be relevant in establishing a right to such fisheries. There is no public right of fishing for salmon in Scotland. Wild salmon, as *feral naturae*, do not belong to any individual person until they are caught.

The right to fish for salmon and sea trout does not depend on the ownership of the adjoining land, but is a separate heritable estate which can be bought, sold or leased. There are exceptions in Orkney and Shetland where some relics of Norse Udal law persist. Here, salmon rights may go with the land. A lease of salmon fishing is protected by the Leases Act of 1449 and it remains in force even if the ownership of the fishing changes. It conveys to the lessee all rights and privileges including an implied right of access to and use of the river bank.

On the other hand, the right to fish for freshwater fish other than salmon belongs to the land adjoining the waters in which the fish are found and cannot be conveyed separately from that land. It may, however, be leased (Leases Act 1449 and Freshwater and Salmon Fisheries (Scotland) Act 1976).

The earliest records of the ownership of fishing occur in the charters awarded by David I and his dynasty when they consolidated their possession of north-east Scotland with the help of the institutions of church and burgh. It may be that David himself incorporated rights to fishings in several charters, although his grant of fishing in the R. Spey in 1124 to the Priory of Urquhart in Moray is the only one which is definitely known. The Priory of Pluscarden (with which that of Urquhart later became joined) was given fishing on the Rivers Findhorn and Spey by Alexander II in 1233. After the Reformation, the fishings belonging to the Priory of Pluscarden passed into the hands of the Duke of Gordon who was the feudal superior of the lands on the lower river. Around 1650, there was a lucrative salmon fishing at Garmouth, at the mouth of the R. Spey, from where 80 to 100 lasts of salmon (143−179 t) were exported annually, although the R. Spey was admitted to be inferior to the R. Don and the R. Dee in its yield. It is interesting to note that the fishings on the lower R. Spey reverted to the ownership of the Crown in 1937.

In common with Inverness, Aberdeen and Stirling, Montrose burgh owned many of the most productive salmon fishings in the neighbourhood, including those in the estuaries of the Rivers North and South Esk. These fishings were gifted to the burgh by the Crown, some probably at its inception in 1140, and others by William I and David II during the 13th and 14th centuries. In addition, William I gifted a fishing on the north bank of the R. North Esk to Arbroath Abbey in its founding Charter of 1178, and a similar fishing on the south bank to Montrose hospital which was founded by the Crown in *c* 1245. On 1 May 1370, David II granted the burgh a charter of feu-ferme on all its properties. This gave the burgh the power to collect and invest in its own common good fund all the burgh rents and petty customs in return for an annual payment to the Crown of £16. These properties included all the burgh lands, crofts and pastures, salt pans, wind and water mills and *cum piscariis infra aqua de Northesk and Southesk in cruis yaris et retibus antiquitus consuetis et pertinentibus ad dictum burgum* − that is, fishings in the North and South Esks in cruives, yairs and nets of old belonging to the burgh. Until the end of the 15th century, the burgh fishings were leased but between then and the end of the 18th century they were gradually all feued off, the burgh deriving only an annual feu duty and entry fees on the succession of an heir (Adams 1985).

The history of ownership of the R. Tweed fishings is rather similar. By the time of the earliest charter in the 12th century, the river from its mouth upstream for a distance of about 16 km had been divided into fisheries, each with its name and limits. The first to be recorded was Hallowstell which was granted to St Cuthbert and his church by the Bishop of Durham before 1122, having previously been gifted by the Crown. Other fisheries were granted about the same time to Kelso and Melrose Abbeys and, strangely enough, to Dunfermline Abbey. The reason for this gift to Dunfermline Abbey may have been because at that time, Scottish kings were resident in the Burgh of Dunfermline.

Berwick burgesses obtained possession of many of the fisheries after 1492 and now exercised a near monopoly which they jealously guarded. In 1505, an Order was passed prohibiting any freeman from letting any fishing on either bank downstream from Horncliffe, unless to another freeman. By the middle of the 18th century, the master coopers of Berwick were owners or lessees of many of the fisheries. They caught and marketed the salmon themselves, employing their own sailing vessels, Berwick smacks (on record from 1756), for the carriage of salmon to London. In 1764, they formed a shipping company which also managed the fisheries. When the shipping side of the business closed in 1872, the Company changed its name from Old Shipping Company to the Berwick Salmon Fisheries Company which operated most of the fisheries near the mouth of the river until they went out of business in 1987.

Many of the charters included salmon fishings in a manner which admitted of no argument. But there were others where the grant was of fishings (without specifying salmon) or of baronial rights without any mention of salmon fishing in particular. These have been taken as including salmon fishing only if the apparent owner has exercised ownership in the fullest possible way, openly, and to the exclusion of others. Meanwhile, until the middle of the 19th century, the Crown stood passively on one side while its regalia were steadily eroded.

In 1832, the management of what remained of the Crown Estate in Scotland (in those days the collection of feudal payments and the ownership of a few farms) was put in the hands of the Commissioners of Woods, Forests and Land Revenues. These Commissioners were already administering the English Crown Properties, and it was not long before they realized the importance of safeguarding what little might remain of the salmon fishing still in Crown ownership. The turning point came in 1849 when the Lord Advocate, acting for the Crown to stop a Mr Gammell fishing for salmon in the sea at Portlethen, successfully maintained that when no evidence could be brought to prove that the Crown had parted with its ownership, the right to fish for salmon in the sea remained part of the regalia of the Crown. Following this case, a survey of salmon fishings and their ownership round the entire coast of Scotland, including the Isles, was undertaken on behalf of the Commissioners of Woods. The result of this investigation showed that the Crown still owned 40% of the fishings round the coast. After allowing for a number of sales, most of which took place at the end of the last century, the position now is that 30% of coastal fishings are Crown property (Hunt 1978). This proportion has been further eroded in recent years by the decision of the Crown to sell considerable numbers of less valuable coastal fishings. A similar review and examination of the Crown's river fishings was completed in the summer of 1990. It will be interesting to see if any ownership questions arise out of this review because in the author's experience it is not unusual to come across angling clubs happily letting salmon fishing to their members together with trout fishing when the salmon rights are in fact still in Crown ownership.

Crown stations are let for a period (up to ten years) by the system of public tendering following advertisement. This method has been used continuously since the end of the last century. One result of tendering can be that the sitting tenant is outbid and, although the sitting tenant was never guaranteed security of tenure, there is every indication that, in the past at least, he was given special consideration. The availability of Crown fishings for let on a regular basis has been extremely valuable as it has provided an entry into the industry, with some security of tenure, for the keen, young, energetic salmon fisherman who has served his apprenticeship. This can only be good for salmon in the long term. Over the years, the system of Crown leases brought together people who handed down from father to son the necessary expertise to manage Scottish salmon fisheries in such a manner that egg deposition has not limited smolt production. They also played a major management role at national level through their membership of district salmon fishery boards, the Association of Scottish District Salmon Fishery Boards and The Salmon Net Fishing Association of Scotland. Their success over the last 150 years is best measured by the stability of the Scottish salmon catch compared with that in many other countries including England and Wales, France, Holland and Germany.

7.2 Ownership in England and Wales

The law relating to the ownership of fishing rights in England and Wales has never been formally incorporated in any legislation. It has evolved over many centuries as a result of court decisions. This is a continuing process and a present-day court decision can overturn previous rulings of a lower court. This new decision will stand until such time as it is superseded by a later decision of a superior court. Nevertheless, the protection of the resource and the control of those who exploit it are enshrined in a number of Acts, the principal one being the Salmon and Freshwater Fisheries Act 1975.

In England and Wales, unlike Scotland, the public has the right to fish in the tidal parts of rivers and the sea except where the Crown or an individual has acquired a private right of fishing or where an enactment has restricted the general right of public fishing. In this context, tidal waters are those parts of the sea which cover the coast below the high water mark of ordinary tides (Millichamp 1987). Prior to the Magna Carta of 1225, it was open to the Crown to exclude the public right of fishing in tidal waters by granting the exclusive right of fishing to a private individual. Since the Magna Carta, however, the Crown has had no right to grant public fisheries to private individuals. Today, the public right of fishing can only be excluded or modified by Act of Parliament. In practice, however, the right to fish in tidal waters in river estuaries may be difficult to exercise because the right to fish does not permit access upon land which is above the high water mark.

All fisheries in non-tidal or inland waters (those parts of a river which lie upstream of the tidal limit in an estuary, such as canals, ponds, pools, ditches and reservoirs) are private, as are the small number in tidal waters granted to private individuals by the Crown before the Magna Carta. One such fishery exists in the lower reaches of the R. Usk. Because this is heritable property, it can be bought and sold in the same way as a private fishery in inland waters. As well as buying the fishing rights, a purchaser will also frequently buy a strip of land along the bank of the river. This gives him control over other activities associated with the land which might otherwise interfere with his fishing.

In order to gain access to fishings, the angler must first identify the owner and then, as in Scotland, obtain the necessary permission to fish. In addition, in England and Wales a licence must be purchased from the appropriate National Rivers Authority (NRA) Region.

7.3 Management in Scotland

Legislation to conserve salmon stocks has a long history in Scotland. The earliest recorded salmon statute dates from the time of Alexander II (1214–1249). In 1318, there was an Act to regulate the use of cruives or traps within rivers. Many Acts were passed in the 15th century prohibiting cruives in tidal estuaries, setting weekly and annual close times, and ensuring the safe downstream passage of smolts.

Salmon were held in such high regard in those days that a meal mill intended for the use of the Burgh of Inverness was closed down in 1474 by order of James III, causing great inconvenience and hardship to the local inhabitants. The reason given was that salmon were being attracted into the lade and killed by the mill wheel. Salmon poaching was also severely punished. Anyone caught stealing salmon from the cruive fishery located on the Ness Islands was liable to be nailed by the ear lobe to a wooden shaft and left for a specific time (*Lugging at the Tron*).

The present form of local administration dates from the Salmon Fisheries (Scotland) Act 1828. The preamble is interesting because it summarizes the statute law affecting salmon fishing for a period of some 400 years:

'Whereas by an Act passed in the Parliament of Scotland in the year One thousand four hundred and twenty-four it was forbidden that any salmon be slain from the Feast of the Assumption of Our Lady until the Feast of St Andrew in winter: and whereas sundry other laws and Acts were made and passed at divers time by the Parliament of Scotland anent the killing of salmon kipper, red and black fish in forbidden time, and the killing and destroying of the fry and smolts of salmon; which Laws and Acts were ratified, confirmed and approved by an Act passed by the said Parliament in the year One thousand six hundred and ninety-six entitled "Act Against Killers of Black Fish and Destroyers of the Fry and Smolts of Salmon": and whereas it is expedient for the preservation of the salmon fisheries in Scotland that the penalties enacted by the said Acts should be augmented and the period of the forbidden time altered and extended, and that sundry other regulations be made' (Ross 1986).

This Act provided for meetings of proprietors to levy rates on a river-by-river basis for the enforcement of the salmon fishing law. The 1828 Act also authorized the appointment of bailiffs with powers of arrest. A fundamental change in the law of evidence was introduced by this Act. It provided that a conviction could be obtained by proof on oath by one or more credible witnesses. The Salmon Fisheries (Scotland) Act 1844 extended the provisions of the earlier Act and for the first time specifically made it an offence for a person not having a legal right or permission from the proprietor of the salmon fishings to take salmon *inter alia* upon any part of the sea within 1 mile (1.6 km) of low water mark. In 1862, a further Salmon Fisheries (Scotland) Act was passed. This might be the basis of present day legislation because it provided for the setting up of district salmon fishery boards. Commissioners were appointed to undertake various duties including the fixing of estuarial limits, the boundaries of the districts, the point in each river for determining the upper and lower proprietors, and the annual close time. They were also required to make regulations with regard to such matters as the observance of the weekly close time, the construction and use of cruives, and the construction of mill dams, lades or water wheels so as to afford reasonable means of access for salmon and the mesh size of nets. These arrangements were further formalized under the Salmon Fisheries (Scotland) Acts, 1863, 1864 and 1868 which, with the 1862 Act, were to be read and construed together as one Act. As such they formed the basis for the administration of salmon fisheries in Scotland until the Salmon Act 1986.

The local administration of salmon fisheries in Scotland is invested in the district salmon fishery boards. As at March 1990, the country was divided into 101 salmon fishery districts on the basis of river catchments but only 59 of those districts had boards in place. A board has a life of three years and consequently all proprietors of salmon fishings meet every third year in order to elect a chairman and, at separate meetings of upper and lower proprietors, to elect members from amongst their numbers to serve on the board. Separate meetings of upper and lower proprietors are held because the constitution of the board calls for a maximum of three upper and three lower proprietors. In addition, a maximum of six representatives (up to three anglers and up to three tenant netsmen) can be co-opted under the provisions of the Salmon Act 1986. The maximum number of persons on a board is therefore 13.

Where there are insufficient proprietors to produce six members in addition to the chairman, the size of the board is restricted to the maximum number of either upper or lower proprietors because there must always be equality of representation as between upper and lower proprietors. Similarly, the numbers of co-opted tenant netsmen and anglers must be the same. They cannot exceed the number of proprietors in the district qualified as upper or lower proprietors, whichever is the smaller.

District salmon fishery boards' powers include the raising of finance by imposing an assessment on each fishery in their district and carrying out works and incurring expenses to protect and improve the fisheries within their district. This area extends seawards for 5.6 km from mean low water springs. Boards can also apply to the Secretary of State for Scotland to make an annual close time order changing the dates of the annual close time or the period within that time when it is permitted to fish for and take salmon by rod and line. Different provisions can be made for different parts of a district. Similarly, on application by the district salmon fishery board, the Secretary of State can also make regulations with respect to the meshes, materials and dimensions of nets used in fishing for or taking salmon and other matters specified in Section 3 of the Salmon Act 1986.

The Salmon and Freshwater Fisheries (Protection) (Scotland) Act 1951 provides that the Secretary of State can require the owner or occupier of any salmon fishery to furnish him with detailed information on the catch of salmon and sea trout made in the fishery. The Act also provides that the resulting statistics can be published to show the catch made by rod and line fishing, by net fishing within estuary limits and by net fishing outwith estuary limits in any salmon fishery district. In practice, this did not occur because catches have been published by method, net and coble, fixed engine and rod and line. There is a saving proviso to ensure that information on the number of fish caught in any one fishery during the previous ten years is not disclosed. The manner in which the statistics can be published was amended by the Salmon Act 1986 to 'in such a manner as may seem to him (the Secretary of State) proper'. Because the Secretary of State has not yet decided whether he wishes to make changes, the manner adopted to summarize and publish the annual catch statistics has not changed.

The law is enforced primarily by water bailiffs appointed by the district salmon fishery boards. Their powers, although fairly wide, are somewhat less than the powers of the police, with whom there is a high degree of co-operation. District salmon fishery boards also work with the river purification boards because of their common interest in controlling pollution. In Scotland, all breaches of the relevant law are reported to the local procurator fiscal in whose jurisdiction the incident is alleged to have occurred. The procurator fiscal will, in the exercise of his discretion, decide whether or not criminal proceedings are in the public interest. Normally, all fishing cases are prosecuted in the local sheriff court by way of summary complaint. The cases are dealt with by sheriffs who are all legally qualified. Sentencing is a matter entirely for the discretion of the individual sheriff but the sentence can be reviewed by the High Court of Justiciary sitting in its appellate capacity.

In Scotland, the annual close times (for netting) begin between 21 August and 14 September and end between 10 and 24 February. The periods of extension for rod fishing end on various dates between 30 September and 30 November. On a few rivers, there is an extension before the close season ends. The earliest date that rod fishing can begin is 11 January in the Helmsdale district. The weekly close time for nets throughout Scotland now extends from 1800 h on Friday until 0600 h the following Monday morning while for rods, fishing is not permitted between midnight on Saturday and midnight on Sunday.

The R. Tweed is not a salmon fishery district except as otherwise provided in the Salmon Act 1986. The river is managed in accordance with the Tweed fisheries acts. The commissioners, who are ultimately responsible for the management of the river, number 81 in total, of whom 43 are appointed by local authorities and the remaining 38 are elected by the proprietors of the fisheries. They are charged under the Tweed Fisheries Acts of 1857, 1859 and 1969 with the general preservation and increase of salmon, sea trout and freshwater fish in the R. Tweed and its tributaries, and in particular with the regulation of fisheries, the removal of nuisances and obstructions and the prevention of illegal fishing. The area of jurisdiction extends 8 km out to sea and includes the coastline between Cockburnspath and Holy Island.

The salmon fishing season on the R. Tweed extends from 15 February to

14 September with extensions for rod fishing from February 1 to 14 and until 30 November. The weekly close time for nets and rods is the same as that in force in Scotland.

The private ownership of salmon fishing rights and the ability to buy and sell these rights together with control of such factors as fishing methods and annual and weekly close times have resulted in the continued exploitation of salmon in the vast majority of Scottish rivers by commercial and recreational fisheries at catch levels which, while varying between years, show no upward or downward trend over the last 100 years.

7.4 Management in England and Wales

The first Act dealing specifically with salmon appears to have been passed in the reign of Edward I (1239–1307). It imposed a penalty on taking salmon at certain times of the year. About 100 years later, in the reign of Edward III (1312–77), other Acts were passed to regulate the construction of weirs, because although these were intended to assist in catching salmon they could also entirely prevent their ascent to the upper nursery areas. These Acts were followed by many others designed for the preservation of salmon fisheries, but in most cases there was little chance of enforcement. Until the Salmon Act of 1865, the enforcement of fishery law was entrusted to conservators or overseers who were appointed by the Justices of the Peace. Their duties were not defined, they were unpaid and their powers were extremely limited. The law became confused and uncertain. The result was that in 1860 a Royal Commission was appointed to investigate the salmon fisheries. Resulting from its findings, a Salmon Fisheries Act was enacted in 1861. This Act places the general superintendence of salmon fisheries throughout England and Wales under the central control of the Home Office. In 1866, Mr Frederick Eden, one of the two original inspectors of fisheries, resigned and in February 1867 the post was taken up by Frank Buckland. In 1896 these duties were transferred to the Board of Trade, and in 1903 to the Board of Agriculture and Fisheries. A further Salmon Fisheries Act was passed in 1865 setting up boards of conservators with jurisdiction over so much of a river or group of rivers as the Secretary of State thought necessary for the proper management of salmon fisheries. This Act of 1865 was important in other ways. In addition to being the first legislation to introduce the principle of the imposition and collection of licence duties, it provided for the appointment to local boards of persons who had a direct interest in the fisheries. While Justices of the Peace still appointed the conservators, there was a provision to appoint as *ex officio* members either local Justices with property interests in the rivers or persons paying above a certain sum in licence duties.

A later Act, in 1873, gave owners and occupiers of fisheries and commercial netsmen representation on these boards. In 1888, county councils were empowered to nominate local government representatives rather than Justices of the Peace.

Under the 1865 Act, 31 fishery districts and boards were defined. By 1894, this number had increased to 53, covering about three quarters of England and Wales. On the recommendation of a Royal Commission on salmon fisheries, which was set up in 1900 under the chairmanship of Lord Elgin, the 1907 Salmon and Freshwater Fisheries Act empowered the Board of Agriculture and Fisheries to constitute and regulate fishery boards and districts by means of Provisional Orders. The 1907 Act also contained an important provision which enabled fishery boards to charge licence duties for freshwater fish. The Salmon and Freshwater Fisheries Act 1923, in addition to consolidating the earlier legislation set out the organizations to be represented on the fishery boards and the number of representatives on each board.

By 1948, the number of effective fishery districts had decreased to 45. The setting up of river boards under the River Boards Act 1948, placed each river system under the charge of an authority responsible for the unified control of salmon, trout and freshwater fisheries, land drainage, and the prevention of river pollution. Under this Act, there were 32 boards, which between them covered the country apart from the Thames and Lee catchment areas and the environs of London. The River Boards Act 1948 provided that the expenses of the boards, so far as they are not defrayed out of other revenues, should be met by precept upon the councils of counties and county boroughs in the river board area. The boards continued to charge for the issue of fishing licences, which often paid the whole of fishery expenditure, and they also had power in certain circumstances to levy contributions on the owners or occupiers of private fisheries. However, their main source of revenue was derived from precepts. The membership of the boards, in normal circumstances, was limited by the Act to a total not exceeding 40, including one member appointed jointly by the Ministry of Agriculture Fisheries and Food to represent local fishery and land drainage interests.

In 1974, ten regional water authorities (RWAs) were set up to manage the salmon, trout, freshwater and eel fisheries within their areas with the specific duty under the Salmon and Freshwater Fisheries Act 1975 to 'maintain, improve and develop' those fisheries. However, only four regions, Northumbrian, North West, South West and Welsh, had significant salmonid fisheries. In addition to managing fisheries, these RWAs had responsibility for water supply and treatment, effluent treatment and disposal, land drainage and flood protection, pollution control, navigation and water-based recreation.

The Salmon and Freshwater Fisheries Act 1975, provides for the basic protection of both juvenile and adult salmon in rivers and coastal waters. The more important provisions in this Act relating to salmon are as follows (Anon 1987c):

(1) All salmon fishing must be licensed,
(2) Close seasons operate for all fisheries; these must have a minimum duration of 92 days for rod fisheries and 153 days for commercial fisheries (242 days for specific types of fixed gear),
(3) Weekly close times apply for commercial fishing; these must be 42 h or more,

(4) A minimum mesh size of 50 mm knot-to-knot applies to all net fisheries, except where approved bye-laws permit the use of smaller mesh sizes,
(5) Certain methods of taking salmon are banned, as is the taking of unclean fish,
(6) The migratory movements of salmon may not be wilfully obstructed in coastal waters or rivers.

This legislation also enables RWAs, by means of bye-laws and Orders approved by central government, to adapt the basic fishery controls and to specify types of gear and modes of operation in their areas.

The Salmon Act 1986, although primarily concerned with Scottish salmon fisheries, introduced additional measures for England and Wales. These measures include:

- A licensee must be present when his net is in operation except in certain specific circumstances,
- It is an offence to handle salmon in circumstances where the person concerned suspects, or it would be reasonable for him to suspect, that the salmon had been taken illegally,
- The use of fixed engines can be authorized by bye-law.

The Water Act 1989, which brought into being the NRA, retained the ten regions and each now has a general manager with managers in fisheries, environment and water quality, flood defence and management, reporting directly to him. The NRA was the first public body with responsibility for fisheries but it preserved a regional and multi-disciplinary approach to the management of the water environment, albeit without the utility functions which were a feature of RWAs. Because there is some central control, the level of co-operation between regions should be greater and the response to similar problems more unified than previously.

The annual and weekly close times for nets in England and Wales are much more variable than in Scotland. Generally, the annual close time begins on 1 September but it can commence as early as 1 August or be delayed until 1 October. There are also major variations in the opening date of the fishing season which can be any time between 31 January and 31 May, depending on the NRA Region in which the fishery is located. In some areas, there are separate annual close times for salmon and migratory trout. Not only can the annual close time differ within a river system but between gears in the same system.

The weekly close time for netting is also varied and, unlike Scotland, need not be confined only to the weekend. Generally, it operates from 0600 h on Saturday until 0600 h on the following Monday morning. However, variations can include ceasing fishing between midnight on Thursday and midnight on Sunday, 0600 h Friday until 0600 h on Monday, 2100 h to 0500 h on Wednesday, Thursday and Friday in addition to 0600 h Saturday to 0600 h the following Monday morning. A nightly close time, lasting from 2000 h until 0400 h may also apply. In some areas, there is no weekly close time for fixed engines. Weekly close times are the same for both salmon and trout.

The annual close time for angling begins between 1 October and 16 December and continues, depending on the NRA area, until some time between 16 January and 14 April. The annual close time for trout fishing differs from that for salmon and generally lasts from the beginning of October until March. There is no weekly close time for angling in England and Wales.

All close seasons and weekly close times are subject to local variations; the exact dates are determined by NRA bye-laws so that an individual river's spawning time can be taken into consideration. Some waters may not be open for the full duration of the statutory season.

From all this, there are marked differences in the management of salmon fisheries in England and Wales compared with Scotland, according to the legal framework under which they are managed. Nevertheless, the Salmon Act 1986 provides, for the first time, new overall measures which are applicable to Scotland, England and Wales.

7.5 Fishing methods

7.5.1 *Netting in Scotland*

Commercial salmon fishing probably dates back at least to the 12th century. The well-known story of the fishery at Pedwell on the R. Tweed, near Norham, first mentioned about 1160, telling how the key of the village church was recovered in the mouth of a large salmon taken in the first draught of the day, implies the use of sweep nets. This story is pre-dated by the tale of Nathalan, Bishop of Aberdeen, who died in 452 AD. Legend has it that for a penance he locked an iron girdle around his loins and threw the key into the Aberdeenshire R. Dee at a spot still known as the 'Key Pool' almost midway between Cambus o' May and Ballater. Thereafter he went to Rome, where in a fish that was brought him by an Italian fisherman he duly found the key (McConnochie 1900). Initially, sweep netting was confined largely to the lowermost reaches of rivers and their estuaries. It was not until the early 19th century that sweep netting also moved on to the open coast as distinct from bays and sea lochs.

The earliest commercial fisheries in Scotland employed a whole variety of nets and traps. The two most favoured were sweep nets (net and coble) and cruives. The latter were weirs built across rivers with a number of gaps into which a trap could be set to catch the fish as they migrated upstream. Cruive and sweep net fisheries frequently operated together. The sweep net fishery harvested the fish whose upstream migration had been impeded by the cruive and had collected in the pools immediately below the obstruction (Plates 6 & 17). Most of the larger Scottish rivers, and many of the smaller streams supported cruive fisheries, with the notable exception of the Aberdeenshire R. Dee. Some rivers supported more than one, and a few continued to operate until about 1960. The cruive on the River Beauly continues to be regularly maintained.

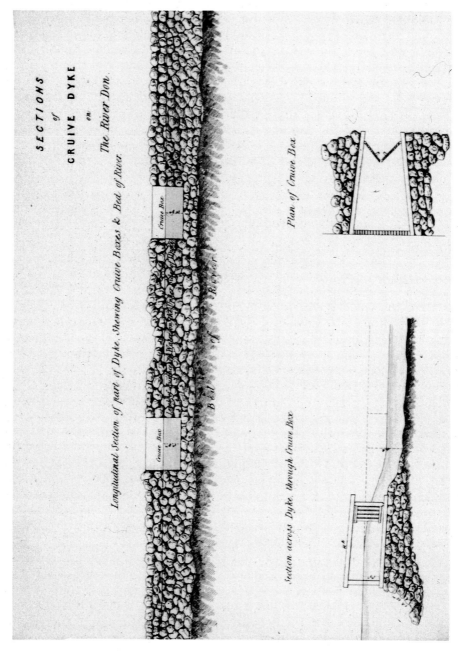

Plate 17 A salmon cruive. (Reproduced from the Report of the Select Committee on Salmon Fisheries, Scotland, 1836).

As well as sweep nets operated from cobles, fixed nets were also permissible in fresh water. Their use in some rivers, particularly the R. Tweed, may have pre-dated net and coble. In the R. Tweed, for example, these fixed nets were called stell and cairn nets. The word *stell* means 'a fish trap, a place for catching fish'. It occurs in the names of some of the earliest fisheries on the R. Tweed, as in Hallowstell and Sandstell. The first sweep net fishery to be recorded was at Hallowstell whose charter was granted to St Cuthbert and his church (that is to the monks of Holy Island) by Ranulf Flambard, Bishop of Durham, before 1122 in the form of the *Haliware stelle*, 'the fishing place of the holy man', a reference to St Cuthbert (Walker 1988).

A stell net is a curtain of netting which is set out by boat from the bank in a semi-circle. A fisherman either holds one end on the bank or fixes it to the shore and the other end is attached to an anchor. Once the net has been set, the boat is positioned upstream of the net near its mid-point and one of the fishermen in the boat grasps the float-line. When fish are felt striking the net or seen within the semi-circle of netting, the anchor is lifted and the net is released and brought ashore. At the same time, the fishermen on the bank pull in their end of the net and as the two ends of the net are gradually brought together, any fish caught come ashore with the net and can be killed.

Cairn type nets were used in many Scottish rivers. They were short lengths of netting set out from piers, usually constructed of boulders, which stretched 2−3 m into the river. Because only one end of the net was fixed, the net swung round in the current and hung parallel to the bank. These nets were particularly effective when rivers were in spate. Any fish then migrating upstream along a bank in the slacker water, when confronted by the cairn tended to turn towards the deeper water and into the net.

Many other types of nets were tailor-made to suit particular locations. They are too numerous to mention. However, some methods (doachs, ladle nets, shoulder nets and yair nets) used to catch salmon in the Galloway R. Dee were unique and deserve special mention. Doachs consisted of a masonry wall joining outcrops of rock, and extending the full width of the river from bank to bank. Three gaps were left to let the water through, but these could be closed to larger fish by inserting removable *hecks*. These were wooden gratings with vertical bars and gaps, a minimum of 75 mm apart. The hecks remained in position except during the weekly close time. During the fishing season excepting the weekly close time, all fish too large to pass between the bars were held back, unless a flood overtopped the wall. It is interesting to note that the legality of this structure has been successfully upheld in more than one court case.

Fish held back by the doachs were caught with ladle, shoulder nets and sweep nets. Ladle nets were outsize landing nets with 6 m shafts and were used to scoop fish out of the pools immediately below the doachs. Shoulder nets were employed in the small pools among the rocks further downstream which were reached by a system of wooden catwalks. The gear consisted of a 7 m shaft with a 2 m wooden crosspiece at the end, kept square by two rope stays. The triangle formed by the crosspiece and stays supported the mouth of a deep bag of netting. Before he made a cast, the fisherman flaked the net on top of the shaft, and rested it on a wooden shoe fitted to his shoulder. The cast was made by shooting the net forward, so that the shaft slid over the shoe and the net fell beyond the fish. The net was then drawn in through the pool, the shaft sliding back over the shoe (King-Webster 1969). This type of netting required great strength and skill, as fishing was only done at night and the foot-holds were often precarious and slippery. Nevertheless, Grimble (1902) mentioned that over a four year-period around 1840, a shoulder net fisherman averaged some 6500 salmon annually.

Before the construction of the hydro-electric station at Tongland, eight yair nets were fished in the estuary of the R. Dee below Kirkcudbright. Each yair consisted of two converging fences or leaders made of stakes interwoven with saplings to form a coarse wicker work. In the apex of the 'V' formed by the converging leaders was a rectangular opening, across the top of which was a platform on which the fisherman sat on a box. The actual net was a deep bag, with shaft and crosspiece, not unlike a larger version of the shoulder net. The fisherman lowered the net into the opening so that it billowed downstream with the water flowing through it. He held a system of lines leading from the end of the bag of netting. When a fish touched the end of the bag, he felt the impact through the lines.

A variety of methods was used to catch salmon upstream from the estuary. Leisters and spears are recorded as being used on both the R. Carron (Stirling) and the R. Spey; 'clips' (gaffs) were in use on the R. Cassley in Sutherland, while a dip-net on the end of a pole was employed on the R. Isla, a tributary of the R. Tay. On bigger rivers like the Spey and Tay, coracles were used to allow the fishermen to get out into midstream. Linns on rivers, where salmon might be exhausted by repeated attempts to jump, were popular places to catch salmon either by hook or gaff and even by hand. Interwoven baskets were strategically set into some falls to catch the salmon as they fell back from an unsuccessful attempt to ascend the fall. The falls on the R. Tummel near Pitlochry were the site of one of the better known basket fisheries as a result of an appeal to the House of Lords against closure. At times, there could be wholesale slaughter of salmon; on the Galloway R. Dee, Viscount Kenmore had the right to 'fence off' the river for eight or ten days, and spears, leisters and even dogs were used to catch the salmon (Coull 1979).

It may come as a surprise to many readers that in the early 1950s, I saw a fishery operated in a somewhat similar manner, on behalf of an estate. The fish immigrating from the sea were allowed to accumulate in the short tidal stretch of river below a loch, whose outlet was regulated by a manual sluice. On the appointed day, the outlet of the river into the sea was closed with several old iron bedsteads, the sluice was shut and as the water drained away the captive salmon were clubbed to death by the laird's employees.

By the 1800s, the owners of river fishing began to realize that considerably higher incomes could be obtained by letting their fishings to anglers. Netsmen, unable to afford the increased rents, were gradually eliminated from the rivers. They went to the coast where in a much harsher environment they developed the fixed engine to catch salmon. Initially, these fixed engines were also fished in river estuaries but, by the mid-1800s, the use of all fixed engines including most trap fisheries was made illegal in rivers and their estuaries. This left sweep netting as the only method of netting permissible inside estuarial limits.

Brief mention should also be made of a drift net fishery which developed off the Scottish coast in 1960. The Reports of the Special Commissioners for the Tweed Fisheries (1875) and the Royal Commission on the Tweed Fisheries (1896) indicated that drift netting was a recognized method of fishing for salmon in the area at that time. The use of drift nets seems to have stopped at about the turn of the century following a decision in the House of Lords in 1900 which declared drift nets to be illegal off the mouth of the R. Tweed. Nevertheless, it remained a lawful method in Scottish waters outside the 4.8 km (3 mile) limit. Drift netting reappeared as a fishing method for salmon in 1960 when a few boats from Northumberland began to fish off the R. Tweed. This method escalated in 1961 and 1962 when boats from Scottish east and north east ports and the Moray Firth became involved. On 15 September 1962, drift net fishing for salmon off Scotland and the R. Tweed was forbidden by an Order made under the Sea Fish Industry Acts 1959 and 1962. The landing of salmon taken by drift net was also prohibited, subject to a provision in favour of those taken off England and Wales under licence issued by a river board or the Minister of Agriculture Fisheries and Food. Drift netting has remained an unlawful method of fishing for salmon off Scotland and the R. Tweed since that time (Anon 1963, 1965).

A single drift net measured *c* 2.5 m deep, 100 m long and had a mesh size around 130 mm. It was suspended from a float line and there was little or no weight along the bottom of the net which was made from hemp twine. Usually some 10 nets were joined together and each boat normally fished two fleets of nets. Most skippers allowed their nets to drift independently and the best results were obtained at night when the wind speed was in excess of Force 3 (Beaufort Scale) and the surface of the sea was disturbed.

Following the banning of drift nets, various attempts were made to modify the gear and its mode of operation in such a way that it would comply with the law. One modification which had fairly widespread application was to attach one end of the net to the shore by lengths of rope in excess of 100 m, thus converting the instrument into a fixed engine in the eye of the law. However, as soon as each modification was perfected, a new Order banning its use was introduced culminating in the banning of any net set to enmesh fish and the carriage of any net made from mono-filament twine, which in the interval had replaced natural fibres in the manufacture of drift nets.

Pelagic long-lines similar to those used in the Faroese fishery proved successful when used in the Moray Firth. On a number of occasions in the 1960s, Danish vessels on passage to the west coast shark fishery landed salmon and sea trout supposedly caught 30–50 km offshore. In addition to immature salmon and sea trout, kelts of both species figured in these catches.

The two main types of fixed engine presently in use are bag and stake nets; neither is permissible inside estuary limits (Fig. 7.1).

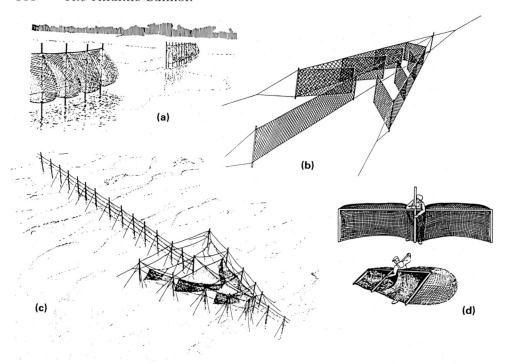

Fig. 7.1 The common types of fixed engine in Scotland: (a) poke net; (b) bag net; (c) stake net; and (d) haaf net.

(1) *Bag net.* The bag net is used on rocky coasts over deep water and consists of two main parts — a leader and a trap. The leader is about 120 m long and is a curtain of netting which stretches out at right angles from the shore to the centre of the trap. The trap is an arrow-shaped structure of netting, supported by wooden poles, into which the fish are guided by the leader, and where they remain until removed by the fisherman. The trap and leader are held in position by a series of floats and anchors (Plate 7). The trap is fished from a specially designed flat-bottomed boat (coble). A recent development has been the use of a double-headed bag net which consists of a trap on either side of the leader so that a complete net has the shape of a letter 'T'.

   Bag nets are normally fished below extreme low water springs on rocky shores, either singly or in connected lines stretching offshore. The topography of the fishing station determines the maximum number of nets that can be fished. This number is rarely achieved because experience has determined the best sites. Limitations are also set by the number which a crew of 5−6 men can handle, usually a maximum of 12 nets.

(2) *Stake net.* Stake nets are known as fly or jumper nets depending on their construction. Their design is similar to bag nets but because they are fished on sandy beaches they can be held in position by stakes driven into the sand. In the case of the fly net, the trap and leader (tiering) are

supported on stakes. No boat is required to fish the net because the catch can be removed when the tide recedes sufficiently to allow a man to walk out on a rope attached to the stakes supporting the fly net. He removes any fish caught with a hand-net (scum-net) through the roof of the trap. In the jumper net, which is normally fished on steeper sloping sandy beaches than fly nets, the leader is allowed to rise and fall with the tide. The net is fished when the tide has receded sufficiently to allow a man to walk into the fish court and remove any fish caught (Plates 8 and 9).

In recent years, bag and stake nets have become more manageable as a result of synthetic materials such as courlene replacing the tarred cotton used in the past for netting and polypropylene replacing hemp and manilla ropes. Similarly, the heavy wooden casks which gave bag nets the required amount of flotation have been replaced by plastic floats. Apart from the use of powered cobles, the manner of fishing has changed little over the last 150 years.

Another two types of fixed net used in the Solway Firth are poke and haaf nets. Poke nets consist of a series of pockets of net in which the fish are trapped as the tide recedes. They are mounted in lines across the tide on rows of poles driven into the mud (Fig. 7.1). Haaf nets are mounted on wooden frames (5 × 1.25 m) with a handle in the middle of the long side. The fisherman stands in the tide with the handle against his shoulder and the net streaming behind him. When he feels a fish strike, he lifts the net to prevent it escaping (Fig. 7.1). Usually five or six men stand in a line outward from the shore and facing the current, their position in the line having been drawn beforehand by lots. When the tide ebbs leaving the innermost net dry, this fisherman moves his stance to the seaward end of the line. This operation is reversed when the tide changes and the water becomes too deep for the outermost fisherman in the line to maintain his position.

The only method of netting presently permissible within estuarial limits is net and coble (sweep netting). A curtain of netting gathered at the centre to form a pocket is released from the stern of a small flat-bottomed boat (coble) as it moves out from the bank downstream. A rope from one end of the net is towed downstream by a fisherman on the bank, while the coble continues to pay out the net in a semi-circle. As soon as is practicable after the net has been paid out, the coble is turned towards the shore, and both ends of the net are pulled on to the beach so that any fish which have been caught in the semi-circle of netting are directed into the pocket and landed. Although this is the standard method of fishing, it may be modified to take account of the topography of the river at particular netting sites (Plate 6). This method of fishing has changed surprisingly little since Thomas Pennant first described it in detail in 1769, and even today some fishermen still prefer oar-power to that of an inboard engine.

During the fishing season, although netting is permissible for 24 hours a day outwith the weekly close time, it is limited by the tide at most stations to a few hours on either side of low water. The fishing routine varies from fishery to fishery. At some stations, shooting and hauling the net are repeated on a regular basis 8—10 times an hour while at other stations shooting the net only occurs when the crew are alerted by a watcher that he sees the signs of fish crossing a ford. Alternatively, fishing may be limited to one shot in the morning and one in the late afternoon, depending on the state of the tide.

As is the case with fixed engines, the minimum mesh size (178 mm all-round when wet) is fixed by statute. In the late 1960s, there was a change-over to nets manufactured from synthetic rather than natural fibres.

7.5.2 *Netting in England and Wales*

In England and Wales, commercial salmon fishing is mainly carried out in coastal waters and in public navigable estuaries where all members of the public are equally entitled to fish and no one has an exclusive right. However, in practice the full effect of the law is modified because in nearly all salmon rivers the fishermen have had to be licensed and RWAs have had power, which many of them have exercised, to limit the number of licences they issued. These powers are now vested in the NRAs. Fixed engines, which are the principle method of fishing for salmon in Scotland, are with four exceptions forbidden in England and Wales. Two of these cover historical privileges and another the use of fixed engines by the NRA. They may also be authorized by bye-laws made by the NRA or a local fisheries committee. Thus, for example, the Northumbrian 'T' nets and the Yorkshire 'T or J' nets are legal fixed engines.

Most methods used are very primitive, especially dip nets in turbid waters. These are simply landing nets and the fisherman pursues individual fish by wading or sometimes from a boat. Designs of nets differ in different areas, as do their names. In the R. Humber area they are known as bow nets or click nets, in the R. Severn as lave nets and on the west coast between the Rivers Lune and Eden as haaf or heave nets.

Lave nets are operated by one man and the net is fastened to a large 'Y'-shaped frame similar in shape to an angler's landing net. The two arms are hinged to the handstaff and kept in position by a wooden spreader which is perforated at one end and notched at the other, so that one arm can be folded down on the other for easier transport. The operator stalks fish which are trapped in pools in an estuary, scooping up any salmon either when stranded or when crossing a sand bar (Fig. 7.2). The haaf or heave nets are similar in construction to those used on the Scottish side of the Solway Firth illustrated in Fig. 7.1.

In the estuaries of the Rivers Severn and Wye and in the eastern and western R. Cleddau in Pembrokeshire, salmon are caught by means of nets operated from a boat anchored across the current. These nets are known as stop nets on the R. Wye and compass nets on the R. Cleddau from their resemblance to a pair of draughtsman's compasses. Each net consists of two arms 3 m long bolted together at one end. These arms are opened to approximately 60°. Across the arms, about 2 m from the bolt, is fixed a lighter pole,

Fig. 7.2 A lave net.

the spreader. Between the arms, a net not more than 2.5 m deep is fixed. When the boat has been moored in a suitable spot, a heavy stone is attached to the arms of the net just above the bolt and the net is opened and fixed. It is then pushed out over the gunwale on the upstream side of the boat and until the ends of the two arms are resting on the bed of the river. Three strings are attached to the net and these pass under the boat and are loosely fixed to the downstream gunwale. When the net is down, the fisherman gathers up these strings until he can feel the net. When a fish strikes, he pulls rapidly on the counter-weighted end of the poles, so elevating the mouth of the net above the water and thus preventing the escape of the fish (Fig. 7.3).

Most of these methods are used locally, but the seine net is used in most estuaries and coastal waters. The seine, draft or draw net used for salmon fishing consists of a wall of plain netting some 200 m long of a depth equal to that of the water where it will be used. It is fished in a similar manner to the net and coble in Scotland (page 101). In this instance, the law dictates that when all the net is shot, the boat returns without pause or delay to the shore whence it set out. The R. Eden in Cumberland is now the only river where, in certain places, seine nets can still be used in the river itself as distinct from in the estuary.

In Carmarthen Bay and in Norfolk, a very simple form of seine net is still used. The net is shot by a man wading into the water whilst his companion holds the shore end. The efficiency of such nets must be low for not only is the depth of water a limiting factor but also the length of net that a man can handle whilst wading.

More salmon are now taken by drift nets than by any other instrument. A drift net is designed to enmesh fish and consists of a wall of netting shot from a boat across the current and allowed to drift freely. One end of the net is attached to a floating buoy or staff and the other to the gunwale of the boat. The head rope is floated and the foot rope leaded to keep the net upright. These nets were used primarily in the Bristol Channel and Solway Firth, and in the estuaries of the Rivers Ribble and Lune. In some cases, the net is shot

Fig. 7.3 A stop or compass net.

Fig. 7.4 A drift net.

across the estuary channel and is allowed to drift freely downstream enmeshing any salmon ascending. In the late 1960s, they were re-introduced off the north-east coast, particularly off Northumbria, and on other migration routes around the coast. Hemp twine has been replaced by nylon and, as a result, these nets are now just as effective in daylight as in darkness. Salmon drift nets are usually several hundred metres long and hang 3 to 5 m deep from the surface. They have mesh sizes between 250 and 260 mm (62.5–65 mm knot-to-knot) (Fig. 7.4).

Local bye-laws for north-east England specify that drift nets used in this area may not be constructed like trammel nets, having two or three parallel sheets of netting, nor include pockets or bags. In addition, they must be shot from a fishing boat manned by not more than four people. The Northumbrian bye-laws also specify that the net must be free to drift with the tide unimpeded by additional weights or attachments.

In the Northumbrian NRA area, a drift net must not be longer than 550 m and the minimum permissible mesh size is 160 mm (40 mm knot-to-knot). In the Yorkshire NRA area the maximum length permitted is 370 m and the minimum mesh size is 204 mm (51 mm knot-to-knot). In the estuary of the Welsh R. Dee, the drift nets are trammels. Trammels consist of either two or three sheets of netting. One sheet, the middle one if there are three, is of standard mesh and the others have much larger mesh usually 600 mm (150 mm knot-to-knot). A fish striking the smaller-mesh net may be held by its gills but it drives a pocket of this netting through the larger mesh (the armouring) and is trapped. The Welsh coracle net is a trammel net which is drifted between two coracles (Fig. 7.5).

In Yorkshire, a number of coastal fixed nets are used and in Northumbria, 'T' nets are in common use. 'T' nets are a type of fixed engine operated close to the shore. They are superficially similar to the Scottish bag and stake nets, and comprise a 'leader' usually about 200 m in length, stretching out from the beach to some form of trap or netting compound. They are of much lighter

Fig. 7.5 Salmon fishing in a Welsh river using a trammel net between two coracles.

construction than Scottish bag nets, and the leader, unlike that in Scottish gears, is designed to enmesh fish. In the Northumbrian 'T' net, the leader forms the upright of the 'T' and two traps with funnel entrances form the crosspiece. Fish swimming parallel to the shore are deflected into the trap or enmeshed in the leader. The minimum mesh size is 152 mm (38 mm knot-to-knot). The net is shot each day and re-set on each tide, and must not be left unattended. Originally, light anchors were used, but following the first 'T' net Order in 1963, heavier anchors were employed and the gear is now regarded as a fixed engine, as defined by the Salmon and Freshwater Fisheries Act 1975 (Fig. 7.6).

In the Yorkshire NRA area, 'T or J' nets are types of gill net. They are any attended or unattended net which consists of plain unarmoured sheets of netting without bags, pockets or monks, not exceeding 360 m in total length measured along the head ropes and which, in the form of a letter 'T' or 'J' is suspended in the water by means of floats and held stationary by anchors and weights. The minimum mesh size is 204 mm (51 mm knot-to-knot). This type of net was introduced into this area in the 1960s. The net is shot at the beginning of each week, and removed at the end of the week. It is classified as a fixed engine.

In the estuaries of the Rivers Severn, Wye and Parrett and in the Bristol Channel, putchers and putts are used to catch salmon. They consist of a wooden framework into which are inserted tiers of trumpet-shaped basket traps. The common form is the putcher, a light wicker basket, open ended with the rods some 5 cm apart, set in its frame so that the mouth faces the

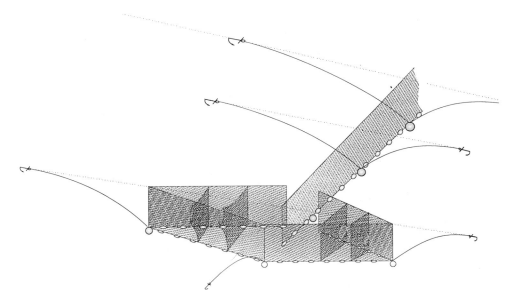

Fig. 7.6 Northumbrian 'T' net.

current, generally the ebb. The salmon, swimming with the current in the turbid water, jams itself head first into the putcher and is trapped (Fig. 7.7). The putt is a much larger and more closely woven conical basket and is a less efficient instrument.

The crib or coop is an extremely efficient instrument but its use has declined in recent years. Good examples can still be found in the Rivers Eden and Derwent in Cumbria. Essentially, it consists of a stone barrier set across the width of the river. Each barrier has a number of gaps into which metal or wooden inscales (a 'V' shaped screen with a gap at the apex) are set, leaving a

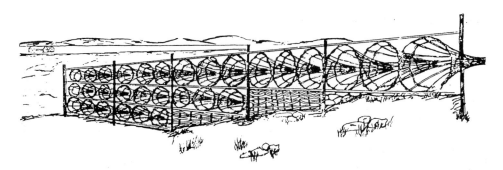

Fig. 7.7 Putchers in the R. Severn.

23 cm gap through which the fish can pass. Their egress to the river above the barrier is stopped by a grating set across the upstream end of the gap and they remain in this trap until removed by the fishermen (Fig. 7.8).

Very few fishing weirs remain but the rights to operate them still exist in some places. Usually, the fishing weir consists of an obstruction across a river channel and can be a natural rock fall or a man-made weir which prevents the upstream movement of migratory fish. When fish try to ascend these obstructions, often without success, they fall back and are trapped in baskets or nets suspended on supports, attached to the face of the obstruction.

The nets described above are the more important presently in use in England and Wales. Others, such as 'P' nets, sling nets, and wade nets have very localized use. A type of fixed engine known as a 'fishing baulk' is unique to Lancashire and probably dates back to monastic times. A similar fixed engine known as a 'garth' still exists at Ravenglass in Cumberland.

As in Scotland, commercial salmon fishing in England and Wales has a long history. This is best exemplified by the primitive nature of many of the methods. Basket weirs are used by very backward fishing tribes in Africa; hand nets must have been used for thousands of years; the seine net was certainly known to the Romans and though historians may still argue about the origin of the coracle net, it is certainly not a modern invention.

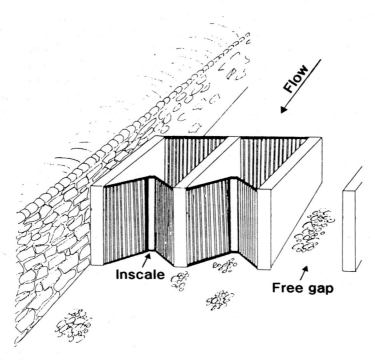

Fig. 7.8 A crib or coop in the R. Eden, Cumbria.

7.5.3 Angling in Scotland

Although Izaak Walton (1593—1683) is recognized as the father of the modern art of angling, the method was actually known and practised in more ancient times. The word *angle*, in its original sense, did not necessarily imply the use of a rod, but meant simply to fish with a hook or angle.

The date when the rod or pole was first used as an attachment to the line and hook is not known, but the first visual proof of its occurrence and use comes from a 20th century BC Egyptian temple drawing (Thomas 1964). The fishing rod was also used in China around the 11th century BC (Trench 1974). However, angling with a rod began not as a means of recreation but as a method of extending the reach of the hand-line fisherman. Nevertheless, ancient Egyptians certainly fished for pleasure, and the story is told that Cleopatra employed divers to place live fish on her hook in order to ensure the success to which a queen was entitled (Russel 1864).

In 1496, the first informative sport fishing report was made. It described a 6 m fish pole hooped with iron, equipped with a horsehair line and several varieties of artificial flies. It was not, however, until the mid-17th century that the pole became more sophisticated and was equipped with an end ring and a reel which was attached to the rod. However, a jointed rod with eyes, line and reel did not come into common use until the end of the 18th century. Nevertheless, some 150 years earlier, Charles Kirby had developed an improved fish hook which is still one of the basic designs (Thomas 1964).

Further development leading to increasing sophistication of the gear continued and has accelerated, particularly during the middle decades of the 20th century. Rods have become shorter. Greenheart and split cane are being replaced by fibreglass, carbon fibre and boron (Plate 10). Wood and brass formerly used in the manufacture of reels have been replaced by light alloys and plastics. The development of fixed spool and multiplier spinning reels has produced equipment which is extremely efficient but so mechanical in its action that much of the art associated with angling, particularly salmon casting, is rapidly disappearing. Nylon and terylene have replaced plaited silk lines. Flies tend to be smaller with hair replacing many of the exotic feathers formerly used. Except for prawns, natural baits, particularly minnows and sprats, are seldom used. They have been replaced by a whole range of ironmongery of all shapes and sizes.

The first recreational anglers were principally of noble status, or commissioned officers in both the army and navy. It was reported that when Lord Nelson lost his arm in the Battle of Tenerife in 1797, a similar misfortune had also befallen one of his young boatswains. When visiting his injured subordinate after the battle, Nelson's chief regret over their shared loss was 'Jack, we're spoilt for fly-fishing' (Russel 1864).

Prior to the first half of the 19th century, fishing for salmon with the rod was largely a local affair. In Scotland, at least, the aim of salmon fishing was primarily to secure something for the pot. From the 1850s, the popularization of rod fishing for salmon made a significant contribution to the rural economy of Scotland. Angling for salmon as a sport for visitors first developed on rivers such as the Tweed which were accessible to the English gentry and merchants who popularized the sport. J. Younger, writing in 1840, reported that John Halliburton, a local man fishing for a living, rented the Dryburgh stretch of the R. Tweed for £15 a year before 1850. Besides the fishing rights, that £15 rent included the ferry boat and its revenues together with 'a cow's grass'. The rent permitted fishing by any recognized method; net, rod, spear. After 1850, he could no longer afford the rent being offered by the English gentry.

During the fishing season, angling is permitted throughout the main river and its tributaries. Proprietors may restrict the number of rods on their beats at any one time and limit the method to fly only or to certain types of lures and baits. However, on some rivers, fly only is statutory during certain times. There may also be statutory bans on the use of natural shrimp and prawn either for part or the whole of the angling season.

7.5.4 Angling in England and Wales

In England and Wales, rod and line fishing is restricted to prevent over-exploitation. Unlike commercial fishing, there is presently no statutory limitation on the number of licences available to anglers. Nor is the rod fishery subject to a weekly close time during the fishing season. In Scotland, on the other hand, licences are not required and angling for salmon is prohibited on a Sunday. Because virtually all rod fisheries in England and Wales are privately owned, owners of such fisheries invariably impose on the anglers restrictions which supplement the constraints imposed by the 1975 Act and the local bye-laws made by each NRA Region. These bye-laws vary widely both within and between river systems. As in Scotland, some fisheries may be fly only while others may also allow spinning and bait fishing. Some fisheries may impose bag limits and some owners may, within the local regulations applying to their fishings, ban the sale of any fish caught.

7.6 Economics

7.6.1 The value of fishing

In 1984, the Tourism and Recreation Research Unit (TRRU) of the University of Edinburgh published the results of a limited study of the Economic Value of Sporting Salmon Fishing in Scotland (Anon 1984c). This study was carried out in 1981–82 for the Department of Agriculture and Fisheries for Scotland. The three areas surveyed were the Kyle of Sutherland, including parts of the Rivers Conon, Cassley, Oykell and Shin, the R. Tay between Perth and Meikleour and the R. Spey between Grantown and Ballindalloch. This survey suggested that the average expenditure per rod/day for those fishing private beats and

Plate 1 The Atlantic salmon.

Plate 2 Young salmon developing inside the egg. The black spots are the developing eyes.

LIVERPOOL JOHN MOORES UNIVERSITY
LEARNING SERVICES

Plate 3 Salmon parr.

Plate 5 Morphie Dyke.

Plate 11 Kinnaber Mill trap.

Plate 12 Automatic fish counting weir at Logie.

Plate 16 Salmon skinned by seals.

Plate 19 Salmon farming in the sea.

those on Fishing Association waters differed between groups but was surprisingly similar within each group.

On private beats, the total amount spent per rod/day was £94–£115. On Fishing Association waters, the expenditure was considerably lower at £43. Both groups of anglers said that accommodation was their major expenditure.

From these figures and estimates of the number of rod/days let on private beats and Association waters, the TRRU calculated that the overall expenditure by salmon anglers in Scotland, including expenditure outside the local area, estimated at 28.7%, was £34 million, with the true value likely to be £22–£46 million. This estimate puts the value of a rod caught salmon at £314–£657 at 1981 costs and £392–£821 at 1987 costs. There were two main sources of error in these estimates. Firstly, there was an error due to sampling in the expenditure survey which varied between 10–50%. (The actual error may have been higher because the sample of anglers was not chosen randomly.) The second possible source of error lay in the estimate of the rod/day totals. The correct figure was possibly higher than the figure of 20% used.

Stansfeld (1989) critically examined the assumptions which the TRRU had made in arriving at these figures, particularly in the context of a straight comparison with the value of the net fishery to the Scottish economy. Stansfeld suggested that these values were too high and that the true value of angling to the Scottish economy was likely to be nearer £22.5 million (£321 per fish) at 1987 costs. Another criticism of the study, which was partially admitted by TRRU, was that the private fisheries involved were among the most expensive in Scotland and outwith the range of most anglers.

In order to determine the management strategy which would be most beneficial to the Scottish economy, Stansfeld (1989) costed three scenarios. These were:

(1) The abolition of angling and, instead, the setting of a TAC for netsmen which would be 80% of the total number of salmon returning to Scottish home waters,
(2) The banning of all netting,
(3) A mix of netting and recreational fishing, as has been the practice for at least 150 years.

Valued on the basis of these three scenarios, the Scottish economy would benefit to the tune of an estimated £21, £26.5 and £27 million respectively.

Because the differences in income in Stansfeld's scenarios were marginal, and because many more benefits, economic, social and biological, would accrue from the presence of a mixed fishery (angling and netting) rather than from only one or the other, Stansfeld suggested that a judicious mix of both types of fishing should be retained.

In the author's view, these figures are simplistic. It is extremely doubtful whether any net fishery could catch 80% of the salmon returning to home waters as Stansfeld suggested. The reasons for this are the present low level of efficiency built into the gear used, the additional restraints of weekly and annual close times, and the fact that in some years more than 50% of the spawning stock may migrate into fresh water after the end of the netting season. Furthermore, there is no evidence that the angling catch would increase by 33% in the absence of netting or knowledge of how a much reduced harvest by all methods might affect smolt production. The degradation of nursery areas if silting is allowed to continue at its present pace may be a more important regulator of smolt production than any realistic rate of exploitation.

In the period immediately following the publication of the paper by Stansfeld (1989), there was a dramatic decline in the price of salmon. This decrease could have a marked effect on the benefits accruing to the Scottish economy from each of the three scenarios examined. However, the recent revival in prices makes his argument once again look sensible.

Whelan and Marsh (1988) estimated that game fishermen in Ireland spent £28 million, of which rather less than 50% came from visiting anglers. A report commissioned by the Scottish Tourist Board (STB) and the Highlands and Islands Development Board (HIDB) to assess the economic importance of salmon angling and netting in Scotland was undertaken by Mackay Consultants and published in October 1989. The estimates of expenditure and income generated were primarily based on ten study rivers/areas. These were the Rivers Thurso, Conon, Orchy, Spey, Dee, Tay, Nith and Tweed and the fishings in the Harris (Hebrides) and Loch Lomond areas. These locations had been deliberately included largely at the request of the two sponsoring bodies for different reasons and may therefore not be representative of Scotland. For instance, as in the previous study, the relatively small proportion of all Scottish salmon rivers surveyed contains the four which are likely to command the highest rents.

Mackay Consultants suggest that the average daily expenditure per angler was £77.23. The main items of expenditure were fishing permits, accommodation, meals and travel. When the average daily expenditure per angler is applied to the estimated number of rod days let in 1988, the gross expenditure in Scotland was about £33.6 million (or £50.4 million when the multiplier is used to take account of additional local benefits). This estimate is within the range of the 1987 estimate provided by the TRRU.

The two main sources of error identified in the TRRU report, particularly the one related to the estimated number of rod/days, would be equally applicable to the two more recent studies. It is interesting to note that actual cost of the permit to fish only amounted to some 30% of the average expenditure per trip and was *c* 10% less than that spent on accommodation and meals. A relatively small proportion of the overall cost of fishing finds its way back to the owner of the fishery.

About 98% of anglers interviewed said that they would return to fish in Scotland and more than 80% had enjoyed their visit. Although 91% thought that the quality of rod fishing could be improved, less than 20% in each instance thought that this could be brought about either by a control or ban on netting or by a reduction in coastal and estuarial netting. About half the hotel respondents rated the trade generated by fishermen as very important, but the other half wrote that it was only of minor importance! Another important point made by Mackay Consultants in their Report was that 'many of the rivers are fished to capacity. Indeed, some of the comments received from anglers point to overfishing'. Thus, there is no room for more anglers. Part of the allure of salmon fishing in Scotland is the exclusivity and many people are willing to pay large sums of money to be able to fish with nobody else in sight.

Mackay Consultants also examined the economic impact of salmon netting in terms of income generated and employment supported. They based their findings both on discussions with members of the The Salmon Net Fishing Association of Scotland, and postal surveys of members of the Association in the same ten-case study areas. Their sample totalled 20 people out of a membership of about 70. In terms of catch, these respondents caught about 50% of the total net catch reported in 1988.

This survey's coverage of the net fishing industry, was considerably less than 30% although it included a number of fisheries whose operators are not members of the Association. Because of this, the basic data are likely to be biased. Nevertheless, Mackay Consultants have estimated the revenue generated by net fishing to be between £1.8 and £3.1 million, depending on what fraction of the total industry the net fisheries surveyed represent.

Although there was a big difference between the revenues and the employment generated by the two methods of fishing, the data are not directly comparable because there are major differences in the manner in which the figures were compiled. While the estimate of the value of angling was based on expenditure, the estimate of the value of netting was based on revenue. No value was calculated for the proceeds accruing from the sale of the angling catch.

Mackay Consultants conclude with the words: 'Both salmon angling and netting make very important contributions to the Scottish economy, with combined output of an estimated £53 million in 1988 and the creation of the equivalent of 3800 full-time jobs'. They list a number of recommendations to achieve growth in the industry and abolition of netting is not one of them.

The average expenditure on salmon angling in the R. Wye has been estimated at more than £800 per fish. This compares with a commercial value of only £36 (Gee & Edwards 1981). Two years later, Radford (1984) investigated the value of four recreational salmon fisheries, the Rivers Wye and Mawddach in Wales and the Rivers Tamar and Lune in England. The total net economic value placed on these four fisheries were £28.72 million, £4.91 million, £15.89 million and £2.40 million respectively, at 1984 costs.

7.6.2 *Marketing*

Traditionally, supplies of salmon were strictly limited. No great effort was required to sell the fish. However, treatment was important and the price depended on careful handling and on getting the fish on ice and to the market as quickly as possible.

Originally, Scottish salmon were salted and packed in barrels. Considerable quantities were exported, particularly to the Baltic ports. There followed a period when salmon were boiled before being packed into kitts (wooden barrels) which were topped up with brown vinegar. If the fish were to be exported to a warm climate, spice was added. Each kitt contained 13 kg (30 pounds). Sufficient salmon to fill 30 kitts was boiled in a kettle at the one time.

The next improvement in the mode of sending salmon to the market, first introduced towards the end of the 18th century and still widely used, was packing them on ice, first in wooden boxes and now largely in polystyrene boxes. This ensured that the fish arrived at the market in first class condition.

In 1960−87, excepting 1976 and 1979, the average price per kg of salmon at Montrose (when corrected for inflation and adjusted to the 1982 level to allow a direct comparison between years), has fluctuated between £3.50 and £6.64 (Stansfeld, pers comm). The current downward trend in the size of wild catches which began in 1976 was initially counterbalanced by a much higher price for individual salmon (£8.09 per kg, adjusted to the 1982 price level). However, this higher price only lasted *c* four years, before declining to a lower level than in most years in 1960−76. In 1986, it fell to the lowest price recorded (Table 7.1). The 1989 value may be even lower than that in 1986 when converted to the 1982 equivalent. However, prices have rallied somewhat in 1990. The main cause of this decline was presumably the increased amount of farmed salmon available on the market and the absence of a corresponding increase in demand.

7.6.3 *Operating costs*

The cost of operating a netting station has increased markedly in recent years. Using the price of a bag net and leader at Montrose as an index, costs increased by *c* 80% from 1978 to 1988. This estimate is almost certainly less than the real increase in running costs because of the continuing increase in labour charges − an increase only partly offset by a decrease in the cost of the twine (courlene) used in the manufacture of nets. Because salmon netting is labour intensive, any increase in wages has a marked effect on the overall running costs of a fishery.

Labour costs probably account for about 50% of the total costs. On the other hand, the average price obtained for 1 kg of salmon from the same centre increased by only 35% between 1978 and 1988 and dropped by 11% in 1989. Because the level of exploitation by nets is relatively low, the netsmen are unable to increase their catch sufficiently to offset the relative decrease in the market price of salmon, particularly as they have to operate within extremely tight statutory regulations. In recent times, these regulations have increased the weekly close time and thus have made it even more difficult to increase catches.

Table 7.1 Prices of Atlantic salmon at Montrose in 1960−89 (£ kg^{-1}).

Year	Average price of wild salmon and grilse	Overall average price of wild salmon and salmon and grilse converted to 1982 prices
1960	0.95	6.13
1961	1.04	6.52
1962	0.75	4.56
1963	0.86	5.07
1964	0.82	4.65
1965	0.84	4.61
1966	0.97	5.09
1967	0.82	4.10
1968	1.01	4.90
1969	1.04	4.76
1970	1.08	4.74
1971	1.28	5.12
1972	1.50	5.60
1973	1.48	5.07
1974	1.65	4.87
1975	2.20	5.25
1976	3.97	8.09
1977	3.77	6.64
1978	3.97	6.46
1979	5.53	7.92
1980	4.83	5.89
1981	4.06	4.41
1982	4.47	4.48
1983	4.18	3.97
1984	4.65	4.23
1985	5.64	4.86
1986	4.21	3.50
1987	5.95	4.76
1988	5.34	
1989	4.39	

Source: Stansfeld (pers. comm)

As a result of all this, net fisheries are ceasing to operate not because of a decline in the numbers of fish caught per trap but because the price of salmon is not keeping pace with operating costs. This situation is very different from that enjoyed in the mid-1840s, soon after bag nets began fishing along the Scottish coast. At that time, a bag net including all its moorings cost £14 compared with £1300 today, a netsman earned 10 shillings (£0.50) per week excluding fish money (one penny per salmon or 3 grilse and one shilling and sixpence per 100 trout). In 1840, salmon fetched 12 p per kg on the London market in July. A weeks wage was, therefore, equivalent to the sale of 10 kg (3 salmon) of salmon compared with 28 kg (9 salmon) in 1989.

Rents paid for salmon net fishings at the beginning of the 18th Century appear to have been relatively high. For example, Hogarth, the tacksman, paid the Duke of Gordon £7000 annually for the right to fish in the estuary of the R. Spey and on the beach on either side of the river mouth. Even with the relatively high rents in the early 1800s, fishings were generally profitable. In 1800, the Netherdon fishing returned a profit of approximately £3000 after meeting all expenses. In 1836, the fishings located towards the northern end of Montrose Bay including Charleton, Kinnaber, Commieston, Kirkside and Woodston but excluding Rockhall commanded a rent of £3800 annually, and it was stated that this rent had remained the same for the last 24 years. The fishing at Rockhall at that time was let for an annual rent of £200. These fishings are no longer let. However, the rateable value of the Rockhall fishing in 1988 was assessed at £5000 and the other fisheries in Montrose Bay at just over £41 000. Even with the relatively high rents in the early 1800s, fishings were generally profitable.

Over a period of 18 years from 1971 to the beginning of 1988, the capital value of rod fisheries increased at a relatively steady rate and their value by the end of the period had increased by a factor of 25. However, during the next 24 months, i.e. to the end of 1989, there was a huge increase which far outstripped the Financial Times SE 100 index. The capital value of a rod-caught salmon increasing on average by 320% to £17 000. (Fig. 7.9). This value has since fallen by some 50%.

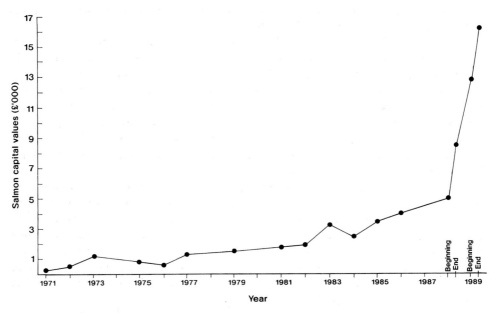

Fig. 7.9 The increase in the capital value of salmon rod fisheries in 1971−90. (*Source*: C. J. Campbell, Strutt and Parker, Edinburgh (pers comm))

7.7 **Summary**

There is no public right of salmon fishing in Scotland, whether in the sea, in estuaries or in rivers, and wild salmon do not belong to any individual person until they are caught. The right to fish for salmon is a separate heritable estate which does not depend on the ownership of the adjoining land, with some exceptions in Orkney and Shetland, and it can be bought, sold or leased. Unlike Scotland, in England and Wales, the public has the right to fish the tidal parts of rivers and the sea except where the Crown or an individual has acquired a private right of fishing or where an enactment has restricted the general right of public fishing.

All fisheries in England and Wales, in non-tidal or inland waters and the small number in tidal waters granted to private individuals by the Crown before the Magna Carta are in private ownership. As in Scotland, these fisheries can be bought and sold.

In addition to obtaining the necessary permission to fish, in England and Wales a licence must be purchased from the appropriate region of the NRA.

Legislation to conserve salmon stocks in Great Britain has a long history, dating back to Alexander II of Scotland and Edward I of England. The local administration of salmon fisheries in Scotland dates from the Salmon Fisheries (Scotland) Acts 1828, 1844 and 1862 to 1868. All but the 1868 Act have been repealed entirely. The 1868 Act was recently modernized under the Salmon Act 1986.

The local administration of salmon fisheries in Scotland is invested in district salmon fishery boards which have a life of three years and consist of a maximum of 13 persons drawn from upper and lower proprietors and representatives of angling and netting interests.

The powers of district salmon fishery boards include the raising of finance and carrying out works and incurring expenses to protect and improve the fisheries within their district. Under the Salmon Act 1986, the boards were given powers to apply to the Secretary of State for Scotland to make regulations by way of Orders relating to such items as the weekly close time and the meshes, materials and dimensions of nets used in fishing for or taking salmon. Prior to the passing of this Act, changes to, for example, the weekly close time would have required primary legislation. In England and Wales RWAs already had the facility to make bye-laws.

In 1974, ten RWAs were set up to replace river authorities and manage the salmon, trout, eel and freshwater fisheries within their areas under the Salmon and Freshwater Fisheries Act 1975. The RWAs also had responsibility for water supply and treatment, effluent treatment and disposal, land drainage and flood protection, pollution control, navigation and water based recreation. The legislation also enabled RWAs, by means of bye-laws and Orders approved by central government, to adapt the basic fishery controls and to specify types of gear and modes of operation in their areas.

The Salmon Act 1986, although primarily concerned with Scottish salmon fisheries, introduced a number of additional measures for England and Wales.

With the enactment of the Water Act 1989, fisheries, land drainage and water quality once again come under the same umbrella and are the responsibility of a new body: the National Rivers Authority (NRA). This is the first national body with responsibility for fisheries which preserves a regional and multi-disciplinary approach to the management of the water environment, albeit without the utility functions, which was a feature of RWAs. The ten regions have been retained.

Commercial salmon fishing in Scotland probably dates back at least to the 12th century. Until the early 1800s, it was confined largely to the lower reaches of rivers and their estuaries. A whole variety of nets and traps were employed but the two most favoured were sweep nets and cruives. By the mid-1800s, the only method of netting permissible inside estuarial limits was net and coble fishing. The netsmen who had been pushed out of the rivers by the owners who accepted the higher rents offered by anglers, responded by developing the fixed engine (bag and stake nets) which could be fished successfully on the coast. Bag and stake nets are still the most common method of catching salmon around the Scottish coast but they are, usually, illegal in England and Wales.

Many types of net have been described as having been used in England and Wales. In 1988, catches taken by 14 different types of net (drift, seine, 'T' net, 'T or J' net, haaf, crib, sling, putcher, lave, trammel, coracle, dip, wade and compass) were reported. However, 95% of the catch was taken in only four gears, drift nets (70%), seine nets (17%), fixed engines (4%) and 'T' nets (3%).

In England and Wales, commercial salmon fishing is mainly carried out in coastal waters and in public navigable channels, where all members of the public are equally entitled to fish. However, the number of active fishermen in each area is limited to the number of licences which the NRA Regions issue.

Two reports, the first describing the economic value of sporting salmon fishing in Scotland and the second covering the economic importance of salmon angling and netting in Scotland, were published in the 1980s. The results of these studies suggested that the overall expenditure by salmon anglers in Scotland, including expenditure outside the local area, may be of the order of £50 million.

In the author's view, more benefits — economic, social and biological — would accrue in the presence of a mixed fishery (angling and netting). Although Mackay Consultants do not plainly state the fact, the implication of their conclusions is that they also come down on the side of a balanced fishery.

In 1840, approximately three salmon had to be caught to pay the weekly wage of a fisherman. In 1989, this number had increased three-fold.

Netsmen have had to operate within extremely tight statutory regulations with permissible gear which has been shown to be inefficient. They have not been able to increase their catch to keep pace with operating costs. As a result, the netting industry has declined; rods, too, have been unable to increase their catch so that the all-method catch has declined.

In the last 20 years, there has been an eighty-fold increase on average in the capital value of rod fisheries.

7.8 Legislation

The principal Acts referred to in Chapter 7 are as follows:

Scotland

Leases Act 1449
Salmon Fisheries (Scotland) Act 1828
Salmon Fisheries (Scotland) Act 1844
Tweed Fisheries Act 1857
Tweed Fisheries Act 1859
Salmon Fisheries (Scotland) Act 1862
Salmon Fisheries (Scotland) Act 1863
Salmon Fisheries (Scotland) Act 1864
Salmon Fisheries (Scotland) Act 1868
Salmon and Freshwater Fisheries (Protection) (Scotland) Act 1951
Sea Fish Industry Act 1959
Sea fish Industry Act 1962
Tweed Fisheries Act 1969
Freshwater and Salmon Fisheries (Scotland) Act 1976
Salmon Act 1986

England and Wales

Salmon Fisheries Act 1861
Salmon Fisheries Act 1865
Salmon Fisheries Act 1873
Salmon and Freshwater Fisheries Act 1907
Salmon and Freshwater Fisheries Act 1923
River Boards Act 1948
Salmon and Freshwater Fisheries Act 1975
Salmon Act 1986
Water Act 1989

CHAPTER 8

THE HIGH SEAS FISHERIES

Until the late 1950s, the harvesting of UK salmon was almost entirely a domestic matter in which privately and publicly owned or leased net and rod fisheries exploited the returning salmon at rates which experience suggested were sustainable. Long- and short-term variation in the age composition of the returning populations and in the timing of the runs seemed to be explicable in terms of environmental changes (George 1982; Shearer 1985a & b, 1988a & b, and 1989). However, by the early 1960s, a number of rapidly expanding high seas fisheries were established outwith UK waters.

8.1 The Greenland fishery

By the early 1900s, there was already a fishery for salmon in the Sisimiut district (Anon 1964) before the warming of the coastal waters and the invasion by cod in the early 1920s. In 1935 and 1936, salmon were reported to have been present at west Greenland in good numbers, especially in the autumn, and 200 fish were caught in the autumn of 1935 (Jensen 1939, 1948). The present fishery dates from 1959 when local Greenland fishermen began setting fixed gill nets from small boats for salmon within the fjords in Maniitsoq district. A rapid expansion of this fishery followed along the coast, and by the mid-1960s, fishing for salmon between July and November was fairly common from Qaqortoq (61°N) to Kangerluk (70°N) with occasional catches coming from Upernavik (73°N). In 1965, Faroese and Norwegian fishermen introduced offshore fishing with free floating drift nets and soon they were joined by fishermen from Greenland and Denmark. The fishing area extended from Kangerluk to Kap Farvel to a distance of 73 km from the baseline.

Initially, the drift nets were made of multi-filament nylon and fished only in darkness. Danish vessels later introduced drift nets of mono-filament nylon which could catch salmon in daylight as well. Nowadays, the standard fishing gear is a light-coloured nylon mono-filament net, with a stretched mesh size of 130−140 mm. A standard unit is about 38 m in length and 4 m deep. The top of each net is supported by styrofoam or plastic floats fixed along the head rope every 2 m and the foot rope is weighted. Flagged buoys with radar reflectors mark intervals along the net. A number of standard units linked together, with swivels to prevent twisting, comprise a drift net. On a large vessel of up to 150 gross registered tonnes with adequate storage capacity for it, a drift net may comprise over 100 such units. Maximum length is normally 3–5 km. In contrast, fixed nets having one end anchored to the shore are seldom longer than five or six units. The nets used for salmon fishing differ from the cheaper type of net used for Arctic charr (*Salvelinus alpinus*) in that they are more transparent acoustically and optically. In 1980, it was not uncommon for a larger vessel to be carrying fishing gear worth $15 000–$20 000 (Kreiberg 1981).

The net is set to fish near the surface but, depending on its specific gravity as determined by floats and fouling, may fish as much as 1–2 m below the surface. The presence of currents causes nets to hang away from the vertical, and reduces the effective working depth. Ice is a serious hazard to drift nets. Since the introduction of nets made of mono-filament nylon, nets are shot throughout the day, and they appear to fish better in poor weather. They are seldom left in the water for longer than a few hours due to the deterioration of the catch and potential losses to seals and amphipods. In 1964 and 1965, 2.8% of the salmon caught exhibited either fresh or healed seal wounds (Shearer & Balmain 1967). While some nets are set close to the head of the fjords, most are set offshore, seldom more than 35 km and more commonly less than 2 km out.

Most of the catch at west Greenland is taken with drift nets with a target mesh size of 140 mm ±5% stretched. The number of drift nets used by each type of boat in 1987 varied considerably. On average, small boats used 40 nets (standard deviation, sd, = 23), each 25 m long, per fishing day whereas bigger boats used 99 nets (sd = 58) per day. Some fixed gill nets were also used, but the number seems to be decreasing each year. In the most recent years, 77–81% of the catch was taken by boats less than 9 m long (Anon 1990a).

Other types of gear which have been fished at Greenland include floating long-lines, the Norwegian Kilenot and the Northumbrian 'T' net, but none has been as successful as drift or gill nets.

Initially, fishing extended from August to November but in recent years the bulk of the catch has been taken within a few weeks of the opening of the season in August.

Reported gill net catches at west Greenland first exceeded 100 t in 1961. By 1964, catches had reached 1539 t and, with the introduction of drift nets and the participation of Faroese and Scandinavian fishermen, catches increased further so that by 1971 the catch had risen to 2689 t. By international agreement, fishing by non-Greenlandic vessels was phased out in 1972–5, but the total catch remained around 2000 t until 1976 when a total allowable catch (TAC) of 1190 t was set and catches thereafter have been regulated (Table 6.5).

Since 1980, small adjustments in mesh size have been made; the current agreed size, fixed in 1982, is 140 mm ±5%. This mesh size was set in an attempt to achieve the same proportions of North American and European salmon in catches as in the exploited population for any opening date of the fishery between 10 August and 1 September. Larger mesh sizes led to catches with a higher proportion of salmon of European origin than in the exploited population while smaller mesh sizes led to catches with a higher proportion of North American origin salmon than in the exploited population. This regulatory action is possible because salmon of North American origin caught at Greenland are generally shorter and lighter than European salmon at the same sea age, and most of the salmon in the exploited stock are of the same sea age group.

The following regulatory scheme was set up in Greenland in order to administer the TAC (Møller Jensen 1988):

(1) No commercial fishing is allowed without a licence issued by the Greenland authorities,
(2) Fish processing plants are not allowed to buy salmon except from licensed fishermen and all plants are obliged to report their accumulated catch daily,
(3) Fishing has to stop immediately when ordered by the authorities. The order is cabled to all fish plants and announced over the radio.

The TAC was divided into two quotas, the so-called: 'free-quota' which all boats and fishermen can fish, and 'the local small boat-quota'. This second quota is allocated on a district basis and only to boats smaller than 9 m. The TAC agreed for the period 1988−1990 was a total of 2520 t, with an annual opening date of 1 August. In addition, the annual catch was not permitted to exceed the annual average (840 t) by more than 10%. In 1988, the TAC was divided into individual 'boat quotas'. This was different from previous years. The aim of the new arrangement was to prolong the fishing season because of the limited freezing capacity in the processing plants and thus improve the quality of their products. On the other hand, such an arrangement with boat quotas could result in a higher discard rate, especially of second class salmon which fetched a lower price.

At present, insufficient data are generally available to assess the number of fish present in the Greenland area in each year and the level of exploitation exerted by the fishery. However, as a result of the international tagging project at west Greenland organized by the International Council for the Exploration of the Sea (ICES), these data are available for 1972. In that year, it was estimated that 1.75 million salmon were present in the west Greenland area at the beginning of the fishery and that the fishery removed 33% of them (Andersen *et al.* 1980).

The salmon fishery in the coastal area of east Greenland is generally restricted, and in some years completely prevented, by drifting polar ice. The reported landings from east Greenland in 1977, 1978, 1985, 1986 and 1988 were 6, 8, 7, 19 and 4 t. In all other years between 1975 and 1988, reported landings were less than 1 t. However, as catches per unit effort in west and east Greenland were comparable, the size of the exploitable stock in the two areas may have been of the same magnitude in the same period.

8.2 The Norwegian sea fishery

Before 1979, the main fishing area for salmon by long-line (see section 8.3) in the Norwegian Sea was to the north of latitude 76°. This fishery was started by Danish vessels in the mid-1960s and they were subsequently joined by Norwegian and, to a lesser extent, Swedish and West German vessels. After the mid-1970s, however, the fishery was again prosecuted almost entirely by Danish vessels as a consequence of a ban on salmon long-lining by Norway. The subsequent extension of the Norwegian fishery limits to 320 km resulted in the fishery by Danish vessels shifting further westwards. This fishery was closed in 1984 under the terms of the North Atlantic Salmon Convention. Fishing took place mainly in the spring and most fish caught were 2SW salmon. Landings peaked at 946 t in 1970 but were more often less than 200 t after the ban on Norwegian vessels in 1976. Although some adults which had been tagged as smolts in Scottish rivers were caught, more than 90% of the tagged fish recaptured were of Norwegian origin.

8.3 The Faroese fishery

Following successful experimental long-lining cruises around the Faroes in April 1968 and 1969 by the research vessel 'Jans Chr Svabo', a small long-line fishery was developed by Faroese fishermen in the coastal waters around the islands. Initially, the area fished was limited and relatively close to the Islands. But within a few years it was extended northwards so that practically no fishing took place south of the Faroes. The season originally lasted from October to June. Annual catches in 1973−8 ranged between 20 and 40 t. These early catches, particularly those taken nearest to the Faroes, were dominated by small, poor quality, 1SW fish.

In 1979, two years after the Faroese 320 km Exclusive Economic Zone (EEZ) was established, the fishery increased substantially and a yield of over 1000 t was attained in 1981. The participation of Danish vessels from 1978−83, an increase in the number of Faroese vessels from nine in 1977 to 44 in 1981 and the extension of the fishing season all contributed to the increased catches (Table 6.2). Furthermore, Faroese vessels were not restricted to the Faroese EEZ until the 1982/83 fishing season. From 1982, the Faroese government agreed a voluntary quota system with a TAC of 750 t in 1982 decreasing to 625 t (25 boats each with a quota of 25 t) in 1983. Although from 1980 onwards it had been a legal requirement to return all fish below 60 cm total length to the water, from the 1984/85 season licences to fish for salmon have included a statement that discards should be handled gently. When necessary, the length of mono-filament nylon attached to the hook (snood) should be cut leaving hooks in situ to minimize trauma. In 1987, the Faroese government extended the power of the Faroese Fisheries Laboratory to close areas to salmon fishing if large numbers of small fish were present in the catches. This latter action was accepted by the North Atlantic Salmon Conservation Organization (NASCO) at their meeting later that year and adopted as one of their regulatory measures. The other regulatory measures agreed at that meeting were that:

(1) The Faroese catch should be controlled in accordance with an effort limitation programme for a trial period of three years (1987−9),
(2) The total nominal catch should not exceed 1790 t and, in any given year, the annual catch should not be 5% more than the annual average (626.5 t),
(3) The number of boats licensed for salmon should not exceed 26,
(4) The salmon fishing season should be limited to 15 January−30 April and 1 November−15 December,
(5) Subject to the maximum annual catch, the total allowable number of fishing days for the salmon fishery in the Faroese zone should be set at 1600 each year.

At the Sixth Annual Meeting of NASCO held in Edinburgh in June 1989, a 2-year measure was agreed allowing the Faroese to catch a maximum of 1100 t over the two years 1990 and 1991, with the flexibility to allow the yearly catch to exceed the annual average by 15%. In effect, the Faroese will be allowed to catch an average of 550 t per annum, a reduction of 47 t (about 8%) compared with the previous agreement.

In total, 22 and 19 licences were actually issued for the 1987/88 and 1988/89 seasons respectively. In the 1987/88 season, some fishing (30% of landings) took place outside the Faroese fishing zone. The fishing in 1988/89 opened on 1 November, as agreed. The closure over Christmas, however, was reduced to 20 December−4 January but the season was closed 15 days earlier to compensate.

After the fishery moved north from the Islands because the fishermen wished to catch the larger fish, most (*c* 90%) of the catch was 2SW fish as opposed to 1SW fish as was formerly the case when the fishery was close to the Faroes. Catches are biased towards females and the proportion of female fish increases with latitude (Table 6.3).

The average vessel is 25 m and is propelled by a 430 horsepower (hp) engine. The ranges in vessel length and engine size, however, are wide, from 15−36 m and 220−1200 hp. The floating long-lines are made up of pins, each of 80 hooks, one line consisting of 10 pins or 800 hooks. The size and type of hooks are No. 3/0 'Mustad' hollow-point. They are tied to a 4.5 m length of mono-filament nylon snood, which is attached by means of a swivel to a main line made of ulstron. The snood is weighted at its mid-point by a barrel lead incorporating another swivel. The gauge of nylon snood from the main line to the barrel lead is 0.5−0.6 mm and between the barrel lead and the hook it is 0.3−0.4 mm. The snoods are mounted 18 m apart, with yellow, 6.5 cm diameter, plastic floats positioned midway between adjacent snoods. Radio-transmitting dhan buoys are attached along the line (Fig. 8.1) at intervals of 4 km. As many as 25 pins may be joined together extending over a distance of

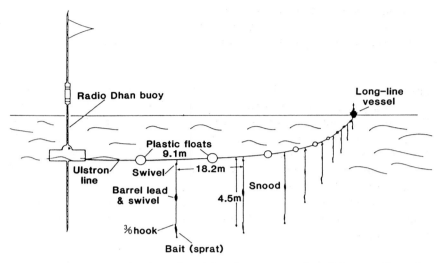

Fig. 8.1 Schematic drawing of a salmon longline. (Reproduced from Mills and Smart 1982)

30 km. The bait consists of sprats which are hooked behind the eyes so that they hang 'tail down'. The line is usually shot at night and hauling begins at first light. It takes 4–5 h to shoot one setting 30 km long and 10–12 h to haul the same. In 1982, a line of ten pins ready for fishing and stowed in a wooden box cost 11 000 Danish kroner (£800) (Mills & Smart 1982). Catches tend to be best in rough weather and at dawn and dusk. Fishing success, as in the 1981/82 season, can be affected by the presence of squid.

In the 1970s and early 1980s, most boats iced their catches, but now all fish are frozen at sea.

8.4 Assessment of the effect of the fisheries at west Greenland and at the Faroes

One of the fundamental requirements necessary to regulate a high seas fishery exploiting fish from many countries is the ability to assess the loss, if not to each river of origin, at least to each country of origin. Except for a handful of rivers, including the R. North Esk, insufficient data are available to allow such an assessment to be made. Consequently, ICES has adopted a general model to estimate the total losses to North American and European stocks for each tonne landed in the Greenland and Faroese fisheries.

8.4.1 West Greenland

The Greenland fishery affects the numbers and composition of the spawning stock and, therefore, the subsequent production of smolts.

The effect on the numbers and weight of fish returning to home waters to spawn will depend on:

(1) The proportion of the original population that visits Greenland (the exploitable stock),
(2) The proportion of the exploitable stock that is caught at Greenland,
(3) The proportion of those fish which avoid capture at Greenland that survives to return to home waters,
(4) The growth of the fish in the time lapse between the mid points of the Greenland and home water fisheries,
(5) The proportion of the returning fish caught in home waters.

Although the effect on smolt production cannot be assessed in the absence of comprehensive data on rivers of origin, the short-term losses (LG) to home water stocks for each tonne landed in the Greenland fishery can be obtained from the following equation:

$$LG = \frac{WH}{WG} \times \frac{S \times 1}{1 - N}$$

where WH = the mean weight of salmon in each smolt year class in home-waters after returning from Greenland; WG = the mean weight of salmon in each smolt year class of European salmon caught at Greenland; S = the survival rate of salmon between Greenland and home waters, the values used being 0.90 and 0.95; and N = non-catch fishing mortality, which is the mortality generated by fishing, e.g. dead fish dropping out of nets, caught in the net and eaten by predators but not recorded as catch. The upper and lower values used are 0.3 and 0.1.

The estimated loss to home water stocks for each tonne of European and North American salmon caught at west Greenland ranges between 1.29−1.79 t and 1.47−2.00 t. The total landing in Greenland, in 1985 for example, was 864 t. Of this total, 415 t were estimated to be North American and 449 t European salmon on the basis of the results of a discriminant analysis of scale characters. From the above equation, 610−830 t and 599−786 t were estimated to have been lost to North American and European stocks respectively. Most (>90%) of this catch would have returned to home waters as 2SW fish if they had not been caught.

Is the Greenland fishery responsible for the present decline in the catch of spring fish in Scotland? Historically, spring fish were absent in the 1800s. In the 20th century, the Scottish all-method catch of salmon taken before 1 May each year (spring fish) peaked in the early 1950s and then began to decline, particularly the 3SW component, before the commencement of the Greenland fishery. The catch of spring fish has shown few signs of recovery even although the catch taken at Greenland has been limited since 1972 by a TAC which is somewhat less than half the catch taken in the years immediately prior to 1972. Furthermore, in neither 1983, 1984 nor 1989 was the TAC caught, the catch in these years being 310 t, 297 t and 337 t compared with a

TAC of 1190 t, 870 t and 900 t (Møller Jensen 1988; Anon 1990a). The Faroese fishery started even later than the Greenland fishery and peaked around 1981 which was even longer after the decline in spring catches began.

In 1987, the Greenland catch was 306 313 fish of which 41% (126 395) were estimated to be of European origin. Based on an instantaneous rate of natural mortality estimated to be 0.01% per month and a time lapse between the mid-point of the Greenland and home water fisheries of some 6 months, the estimated number of fish taken by the fishery which might have survived back to European home waters is:

$$126\,385 - 7360 = 119\,025 \text{ fish.}$$

These fish represent the estimated loss to home water fisheries and spawning stocks in all European countries contributing to the Greenland fishery in 1987. Of the European countries contributing to this fishery, Scotland is thought to provide the largest share, say 50% or an estimated 59 513 fish. At an estimated exploitation rate of 20%, losses to Scottish commercial fisheries would be 11 903 fish. If all the remainder return to fresh water before the end of the rod fishing season, the estimated loss to all rod fisheries, at an exploitation rate of 10%, would be 4761 fish. This estimated loss to the angling catches is unlikely to be equally distributed among all Scottish rivers. Nevertheless, the loss to the R. Dee rod and line catch, because of its above average dependence on MSW fish, may be greater than most, and in this example could be some 750–1000 fish. Even doubling that number to account for losses to the Faroese fishery in recent years, it is unlikely that these fisheries alone are responsible for the decline in spring catches in Scotland from the high plateau prior to 1964.

8.4.2 The Faroes

Basically, the model is the same as that used to assess losses to home water stocks resulting from the west Greenland fishery. But in this instance, the losses are estimated separately for each smolt age class returning to home waters after spending different periods in the sea (Anon 1981, 1984a and 1986b). The equation used to estimate the total short term losses is described in Appendix B.

Subject to the limitations imposed by the estimation of input parameters, the current estimate of home water losses for every tonne intercepted by the Faroes fishery is 1.59 t (Anon 1987a).

Relative losses to home waters will depend on the life history stage at which the fish were harvested in the interception fishery. With regard to the Faroese fishery, the highest relative losses occur when harvesting young fish which would have matured one year later, and losses are least when harvesting older fish which would have matured in the same year.

On the basis of the differences in maturity status and sea age composition of the catches in the Faroese and Greenland fisheries, it has been concluded that the Faroese fishery does not harvest significant numbers of salmon that would otherwise be available subsequently to the west Greenland fishery. However, the Faroese fishery may be harvesting salmon on their return migration from west Greenland to European rivers. Furthermore, the Faroese fishery may be affecting spawning stocks which contribute to both the Faroese and Greenland fisheries (Anon 1984a).

8.5 The Irish fishery

The drift-net fishery to the west of Ireland is also known, on the basis of the recapture as adults of fish tagged as smolts, to intercept some fish (mainly 1SW) returning to UK rivers. This fishery tends to operate only in the late spring and early summer and 1SW fish make up more than 90% of the catch. In the 1960s, there was a massive increase in the catching power of the Irish drift net fishery, largely due to the introduction of nets made from synthetic twine. The use of the original mono-filament nets was restricted in 1984 and banned in 1985. As a result of the introduction of mono-filament nets, the proportion of the total catch taken by drift net in 1951−83 increased from 16% in 1961 to 83% in 1983 (Browne 1986) and the Irish wild salmon catch in 1988 was the largest in the North Atlantic area (Anon 1989c). Although this fishery is restricted to Irish coastal waters out to 19 km from the baseline, there is evidence that in recent years some fishing may have taken place outside this area. As a result, the proportion of non-native fish in the catch may have increased. Because the statistics of the fishery are unreliable and catches cannot be divided between rivers of origin, the impact of this fishery on home water stocks cannot be assessed.

To achieve effective management of the Irish salmon fishery, the Salmon Review Group (Anon 1987b), set up by the Irish government, made 11 recommendations including the restriction of fishing boat lengths to 12 m and the prohibition of drift netting for salmon outside 14.5 km from baseline in 1989, outside 9.5 km by 1990 and subsequently outside 5 km. None of these recommendations has yet been put into effect.

8.6 Fishery for salmon in international waters

During 1989 and 1990, there has been a number of reports of fishing for salmon in international waters by vessels that were registered in countries that are not Parties to the NASCO convention. The reports indicated that some of the vessels were skippered, however, by Danish nationals and it appears that the re-flagging was a mechanism to avoid the provisions of the NASCO convention. The re-flagging so far has been to Poland and Panama. At least seven vessels operated and it was estimated that up to 630 t of salmon may have been taken in 1989−90.

8.7 Summary

In the 1960s, net fisheries for salmon were developed off the west coast of Greenland and a pelagic long-line fishery in the Faroese area. Marked fish caught in these fisheries were found to have originated in most salmon-producing countries, but few North American fish were identified in the Faroese catches.

From about 1965−1980, the main fishing area for salmon by long-line in the Norwegian sea was to the north of latitude 76°N. Landings of mainly 2SW fish peaked at 946 t in 1970. More than 90% of the tagged fish recaptured were of Norwegian origin. This fishery was closed in 1984 under the terms of the North Atlantic Salmon Convention.

Following successful experimental long-lining cruises around the Faroes in the late 1960s, a fishery was developed by Faroese fishermen, initially limited to the coastal waters around the Faroes. In recent years, this fishery has extended northward to the limit of the Faroese 320 km EEZ. At present, the fishing season extends from 15 January−30 April and 1 November−15 December.

The fishery, catching mostly 2SW fish, peaked at 1025 t in 1982, since when catches have been regulated. The quota was reduced from 750 t to 597 t in 1982−89. In most years, the fishery has failed to catch its quota. Other restricted measures imposed on the fishery have included a limitation on the number of licences, the length of the fishing season and the amount of fishing effort.

In most years, over 90% of the salmon caught in these two fisheries would have otherwise returned to their home rivers as 2SW fish. Female fish predominate in both fisheries. In the Faroese fishery, the proportion of females caught increases with latitude. While very few potential grilse are caught in the Greenland fishery, the Faroese catch consists partly of this age group and the fish originate in European rivers. One sea-winter fish are more common in catches taken in areas where the sea surface temperature exceeds 4°C. Although pre-grilse frequent the Faroese area, they do not appear in landings because it is illegal to land fish of 60 cm or less (total length).

At west Greenland, most of the catch is taken with drift nets which have a target mesh size of 140 mm when stretched. The number of fixed gill nets used seems to be decreasing each year. The bulk of the catch is now normally taken within a few weeks of the opening date in early August. The catch exceeded 100 t for the first time in 1961 and peaked at 2689 t in 1971. Although fishing by non-Greenlandic boats was phased out in 1972−5, the total catch remained around 2000 t until 1976 when a TAC of 1190 t was set. Thereafter, catches have been regulated and in the period 1976−89 between 297 t and 1395 t were taken.

In years when January and August in the north-west Atlantic are colder, fewer salmon may move into the west Greenland area than in warmer years.

Salmon caught at west Greenland average 3 kg in weight. Corresponding values for fish caught in the Faroese fishery are about 3.5 kg. European salmon caught at west Greenland are on average heavier and longer than North American fish of the same sea age.

Estimates of the loss to home-water stocks for each tonne of European salmon caught at west Greenland range between 1.29 and 1.79 t. For the Faroese fishery the estimated loss to home-water stocks per tonne of fish caught at Faroes is 1.59 t.

A small fishery, catching less than 10 t in most years, operates in the coastal area of East Greenland.

The northern marine fisheries are unlikely to be the main cause of the present decline in the catch of spring fish in Scotland.

On the basis of the recapture as adults of fish tagged as smolts, the drift net fishery to the west of Ireland is known to intercept some fish (mainly 1SW) returning to UK rivers.

A small number of vessels have been re-flagged to Poland and Panama in order to avoid the provisions of the NASCO Convention. These vessels may have caught 630 t of salmon in the North-East Atlantic Commission area in 1989−90.

CHAPTER 9
MIGRATION AND EXPLOITATION

9.1 Migration in coastal waters

Although our knowledge of the migrational pathways of salmon in the open sea is extremely scanty, the coastal movements, particularly around the Scottish coast, have been studied in some detail (Menzies 1937, 1938a, b & c and 1949). Calderwood (1940) summarized the earlier experiments and Pyefinch & Woodward (1955) described a number of experiments in the area of Montrose Bay in the late 1940s. In addition to providing information on the coastal movements of salmon, these experiments indicated where some of the salmon belonging to particular rivers were being harvested and the rate at which salmon migrated along the coast. A further series of coastal tagging experiments was begun in 1952, starting at Altens, just south of Aberdeen, and finishing in 1988 at Berriedale, south of Wick. In various years during that period, salmon (mainly grilse) caught in bag nets at 13 sites around the Scottish coast were tagged before release (Fig. 9.1). These experiments provided further information on the coastal movements of fish and for the first time allowed estimates of the levels of exploitation by the various gears (Shearer 1986a).

Fish tagged at Fascadale (Argyllshire) were recaptured at points on the coast between the Solway Firth and Montrose (also in Mull, Arran, Islay and Skye) and in the Rivers Awe, Lochy, Spean, Shiel, Morar, Inverie, Dee (Aberdeenshire) and Tay (Fig. 9.2).

Grilse tagged on the Scottish north-west coast at Enard and Badentarbat Bays were also recaptured on much of the coast between Ardnamurchan and Aberdeen (also in Mull, Skye and Lewis) and in the Rivers Carron, Luing, Balgy, Ewe, Gruinard, Broom, Ullapool, Garvie, Inver, Laxford, Naver, Halladale, Thurso, Shin, Spey, Dee and North Esk. Each year, recaptures in freshwater of fish tagged at these northern and western sites tended to be single fish at each site, so that there was little evidence of stocks from particular rivers making major contributions to this fishery (Fig. 9.3).

Recaptures from the tagging sites on the north coast also occurred all round the Scottish coast from the R. Ayr on the west to the R. Tay on the east coast, in the Rivers Laxay, Grimersta and Barvas on the Isle of Lewis and the Rivers Laxford, Dionard, Hope, Thurso, Berriedale, Shin, Beauly, Spey, Dee, North Esk and Tay on the mainland (Fig. 9.4); and two fish were caught in Ireland. However, in 1977, 1978 and 1979, 40–56% of the recaptures were taken in the Rivers Naver and Halladale, the two major salmon-producing rivers nearest

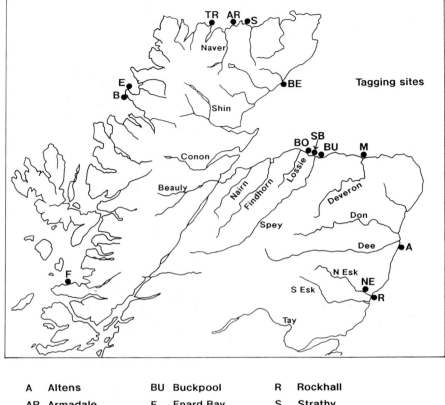

Fig. 9.1 Sites where salmon were tagged in 1952–88.

A	Altens	BU	Buckpool	R	Rockhall
AR	Armadale	E	Enard Bay	S	Strathy
B	Badentarbat	F	Fascadale	SB	Spey Bay
BE	Berriedale	M	Macduff	TR	Talmine and Rabbit Island
BO	Boars Head	NE	North Esk		

to the tagging site. Two fish tagged at Strathy in June and July were recaptured at Enard Bay nine and three days later (Fig. 9.3). Both fish were subsequently recaptured at Armadale (Fig. 9.4) and at Barvas (Isle of Lewis) one to two weeks later. These results show fish do not always travel continuously in the direction suggested by their initial recapture site.

Grilse tagged on the north east coast at Berriedale in 1984–88 were recaptured on the coast between Enard Bay and Lossiemouth (as well as in Skye) and in the Rivers Halladale, Thurso, Berriedale, Helmsdale, Brora, Shin, Oykel, Alness, Findhorn, North Esk, Tay and Tweed. Apart from the Rivers Berriedale, Helmsdale and Brora, recaptures at most other freshwater sites were of single fish, but at Berriedale there was a relatively high recapture rate in the nets from which the tagged fish had been released (Fig. 9.5).

Recaptures of fish tagged in the Moray Firth occurred on the east coast

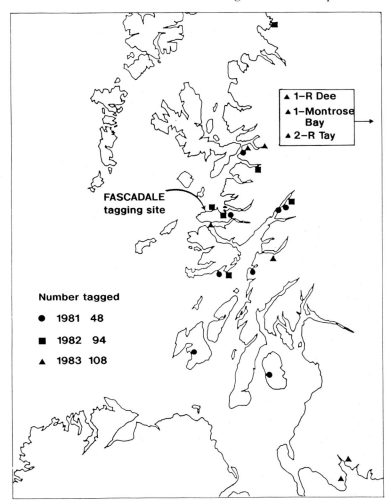

Fig. 9.2 Recapture sites of salmon tagged from coastal nets on the west coast in 1981–3.

between Helmsdale and Montrose, and in the Rivers Helmsdale, Conon, Findhorn, Lossie, Spey, Deveron, Ythan, Don, Dee, North Esk, South Esk, Tay, Tweed and Aln (Northumberland). Each year, between 19 and 44% of the recaptures were caught in the Rivers Spey and Deveron, the two major salmon-producing rivers nearest to the tagging sites (Fig. 9.6).

From fish tagged at Altens near Aberdeen, 30% of the recaptures were caught in the R. Dee, and 23–43% of the recaptures from fish tagged at Rockhall near Montrose in 1954, 1955, 1977 and 1978 were caught in the Rivers North and South Esk. Fish from Altens were also caught in the Rivers Don and Tay and from Rockhall also in the Rivers Nairn, Spey, Don, Dee, Tay and Tweed (Fig. 9.7).

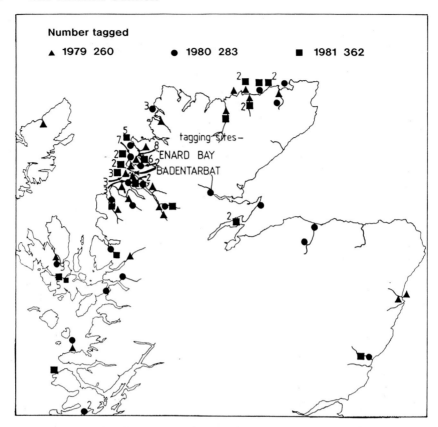

Fig. 9.3 Recapture sites of salmon tagged from coastal nets on the north-west coast of Scotland in 1979–81.

During the 25 years between the first and last experiment, these was no significant change in the pattern of either the coastal or river recapture sites. This suggests that the landfall of the fish and their subsequent dispersal have not altered.

On the assumption that all the fish originated from north west of the British Isles, the main direction of movement differed between tagging areas. Generally, the results indicate that fish move north on the west coast, east along the north coast, south on the north east coast, east along the shore of the Moray Firth, and north on the east coast. However at Altens there was no preferred direction of movement. This pattern presupposes that the fish must move south on the east coast outwith the range of the coastal fisheries before turning north again (Fig. 9.8). This suggestion is supported by the results obtained from tagging experiments carried out in the Northumbrian drift net fishery where most (>90%) recaptures occurred north of the tagging site (Potter & Swain 1982).

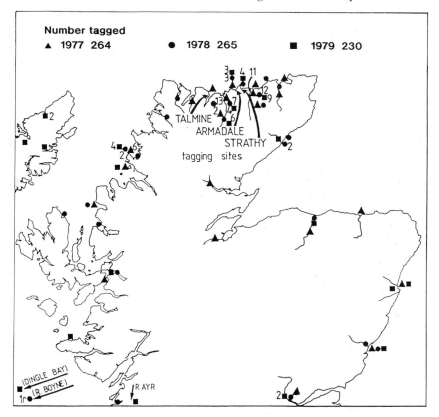

Fig. 9.4 Recapture sites of salmon tagged from coastal nets on the north coast of Scotland in 1977–9.

9.2 Exploitation rates

Because the fisheries at which the fish were tagged were exploiting fish originating from more than one river system, they are referred to as mixed-stock fisheries. In this context, exploitation rate has been taken to mean the proportion of the assumed available tagged fish which was removed by each of the gears. About 90% of recaptures by nets occurred within 15 days of tagging. In most years, 75% of these recaptures were taken within three days. This indicates that fish migrating along the coast are available to the shore-based net fishery for up to 15 days. Tracking experiments show that salmon do not always hug the coast but they may move 1–2 km offshore. They could migrate outwith the netting zone for at least part of their journey back to their home river and therefore be available to the nets for a relatively much shorter period.

The rate of exploitation at each site has been calculated from the number of fish tagged in each area and the number of fish subsequently recaptured by net

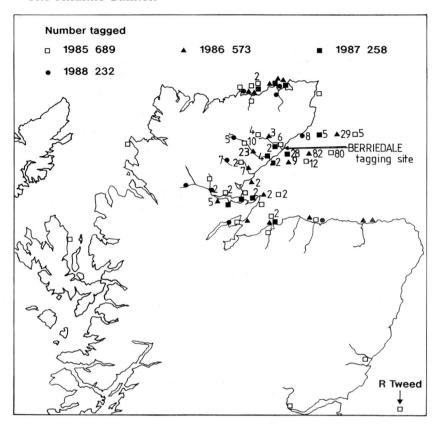

Fig. 9.5 Recapture sites of salmon tagged from coastal nets on the north-east coast in 1984−8.

and rod and line fisheries. Fish caught by fixed engine or net and coble were subsequently unavailable to rod and line fishermen.

The experiments showed that the level of exploitation varied greatly between different areas of Scotland, but tended to remain much the same between years at individual sites (Table 9.1). Whereas the mean exploitation rate by all nets of grilse tagged on the east coast was 55%, the corresponding values for the Moray Firth, north-east, north, north-west and west coasts were 20%, 19%, 16%, 9% and 7% respectively. As might have been expected, the level of exploitation around the coast was directly correlated with the densities of nets that were fishing.

In the case of grilse, the fixed engine fisheries usually caught a significantly higher proportion ($P < 0.001$) than the net and coble fisheries, except on the north coast. This finding has led people to assume that some grilse did not enter rivers to spawn. This is unlikely to be true because all fish recaptured in the year following tagging had spawned in the intervening period.

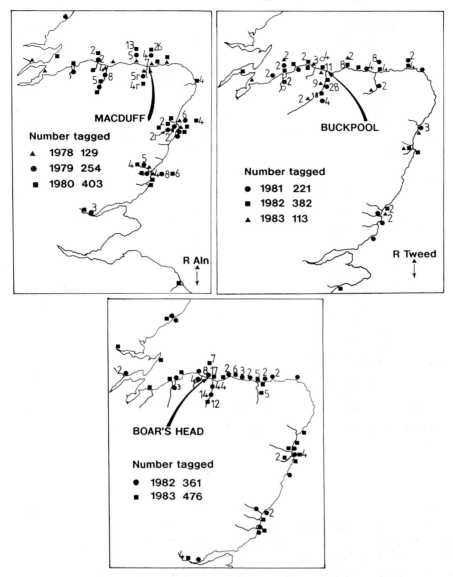

Fig. 9.6 Recapture sites of salmon tagged from coastal nets in the Moray Firth in 1978–83.

Because the coastal tagging experiments coincided with the period of maximum fishing effort, the exploitation rates reported for coastal fisheries are minimum estimates of the maximum exploitation. However, local exploitation rates within river systems with heavily exploited stocks will be underestimated by averaging their data with rivers with lower exploitation rates. Moreover, the figures refer to exploitation rates within the fishing season, and a variable

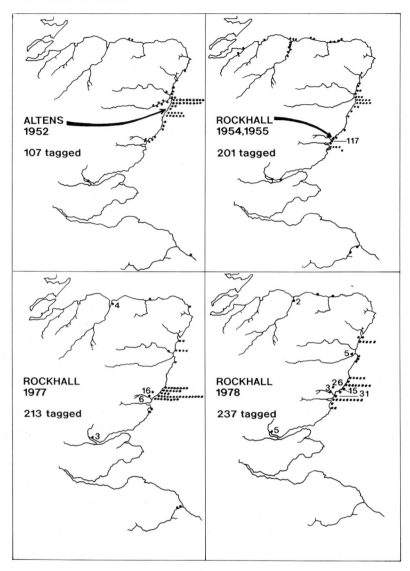

Fig. 9.7 Recapture sites of salmon tagged from coastal nets on the east coast in 1952–83.

proportion of the total stock is not available for catching because it enters rivers during the close season. Nevertheless, in only one area (east) did the estimated level of exploitation exceed 50%. These data refer to grilse, but on the few occasions when salmon and grilse were tagged in the same year at the same site (Moray Firth 1983, east coast 1954, 1955, 1978), the corresponding total exploitation rates were not significantly different from each other (Tables 9.1 & 9.2).

Fig. 9.8 Assumed direction of movement of salmon around the Scottish coast.

During salmon tracking experiments in Montrose bay, four of the salmon encountered salmon nets while moving along the coast, or into the river, and in each case successfully negotiated the hazard. Two fish, in particular, repeatedly avoided capture by both bag and stake nets as they swam to and fro along the beaches to the north and south of the R. North Esk.

Sometimes the fish swam to the seaward end of the line of nets and then rounded it. In many cases, however, the fish swam between the leader and the cleek on one side, around the end of the leader, and then out between the leader and the cleek on the opposite side (Fig. 9.9). Other fish passed around the seaward end of a bag, then swam between the bag and the leader of the next net in the same line.

From the ease with which salmon avoided bag and stake nets it was concluded that the higher recapture rate on the east coast reflects the frequency with which salmon met salmon nets, rather than the efficiency of the individual nets *per se* (Hawkins *et al.* 1979a & b).

The proportion of tagged fish taken by rod was generally no greater than 5%, suggesting that on the north, north-east and east coasts, including the

Table 9.1 Expoitation rates (U) on one sea-winter salmon tagged at coastal netting stations in 1952–88, expressed as percentages.

Area	Year	Number tagged	Fixed engine		Net and coble		Rod and line	
			U	CL[1]	U	CL[1]	U	CL[1]
West coast	1981	48	2	—	4	—	14	—
	1982	94	4	—	—	—	3	—
	1983	108	4	—	5	—	3	—
	Overall	250	4	± 2	3	—	5	± 3
North-west coast	1979	260	7	± 3	2	—	4	± 3
	1980	283	8	± 3	4	± 2	2	—
	1981	362	6	± 2	4	± 2	2	—
	Overall	905	6	± 2	3	± 1	2	± 1
North coast	1977	264	6	± 3	10	± 4	4	± 2
	1978	265	5	± 3	10	± 4	3	—
	1979	230	7	± 3	10	± 4	4	± 3
	Overall	759	6	± 2	10	± 2	3	± 1
North-east coast	1985	689	16	± 3	3	± 1	3	± 1
	1986	573	23	± 3	2	± 1	8	± 2
	1987	258	16	± 4	1	± 1	3	± 2
	1988	232	5	± 3	0	± 1	8	± 4
	Overall	1752	17	± 2	2	± 1	5	± 1
Moray Firth	1978	129	13	± 6	4	—	2	—
	1979	254	11	± 4	9	± 4	4	± 3
	1980	403	15	± 4	4	± 2	3	± 2
	1981	221	13	± 4	15	± 5	5	± 3
	1982	743	9	± 2	15	± 3	6	± 2
	1983	398	7	± 2	8	± 3	5	± 2
	Overall	2148	11	± 1	9	± 1	5	± 1
East coast	1954	286	28	± 5	35	± 7	—	—
	1955	105	27	± 8	21	± 9	—	—
	1977	207	23	± 6	19	± 6	2	—
	1978	188	37	± 7	25	± 8	1	—
	Overall	786	29	± 3	26	± 4	1	—

[1] 95% Confidence limits

Moray Firth, angling removed a relatively small proportion of the fish escaping the nets. However, on the west coast where the density of nets at each fixed engine station was lower and the stations more widely spaced, the exploitation rate by rods was similar to or greater than that of either of the netting methods. In addition, the western rivers are generally narrower so that a greater proportion of the water is accessible to the angler.

The coastal tagging experiments mainly involved mixed stock fisheries and did not allow any assessment of the exploitation rate for the stock migrating

Table 9.2 Exploitation rates (U) on multi sea-winter salmon tagged at coastal netting stations in 1952−83, expressed as percentages.

Area	Year	Number tagged	Fixed engine		Net and coble		Rod and line	
			U	CL[1]	U	CL[1]	U	CL[1]
Moray Firth	1983	191	8 ±	4	6 ±	4	4	—
East Coast	1952	127	29 ±	8	16 ±	7	8	—
	1954	209	24 ±	6	36 ±	7	4	—
	1955	96	32 ±	9	25 ±	10	—	—
	1978	49	27 ±	12	28 ±	15	—	—
	Overall	481	27 ±	4	28 ±	5	4 ±	2

[1] 95% Confidence limits

into individual rivers. Data on exploitation rates by single stock fisheries came from tagging adult salmon at Kinnaber Mill trap (R. North Esk) and at the mouth of the R. Spey (Plate 18).

Exploitation rates for the net and coble fishery in the R. North Esk in 1976−88 were 35−62% for grilse and 30−63% for salmon (Table 9.3). After 1981, the fish counter at Logie allowed comparable rates of exploitation to be calculated on the total spawning stock rather than on the fraction which was available during the fishing season, which was <50% in some years. These rates in 1981−87 were 15−40% for grilse and 29−59% for salmon (Table 9.4). These two exploitation rates fluctuated independently. Accurate estimates of rod exploitation levels in this river are not available mainly because of either

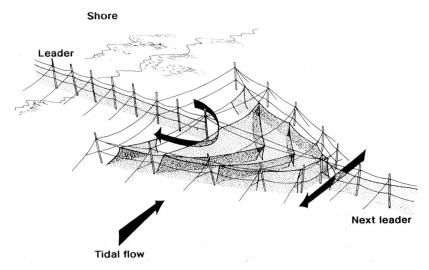

Fig. 9.9 Avoidance of salmon nets by salmon.

Plate 18 Tagging adult salmon with a conventional tag.

Table 9.3 Estimated exploitation rates on one and multi sea-winter salmon in the R. North Esk net and coble fishery in 1976—88, expressed as percentages.

Year	One sea-winter salmon	Multi sea-winter salmon
1976	52	55
1977	50	43
1978	44	51
1979	42	46
1980	39	39
1981	50	62
1982	50	63
1983	53	39
1984	62	44
1985	35	53
1986	51	30
1988	46	30

Table 9.4 Exploitation rates on the total R. North Esk stock of one and multi sea-winter salmon by the in-river net and coble fishery in 1981–87, expressed as percentages.

Year	One sea-winter salmon	Multi sea-winter salmon
1981	23	59
1982	30	49
1983	15	31
1984	28	42
1985	23	35
1986	40	29
1987	29	38

As derived from Logie counter data and includes the fish migrating during the annual close time

the lack of meaningful rod catch data or information on the effort put into angling.

The estimated levels of exploitation by the rod and line fisheries in the R. Spey on salmon and grilse combined were similar in 1983, 1984 and 1985 at 7 ± 3%, 5 ± 2% and 8 ± 3% (Shearer 1988a & c). However, the level at which the net and coble fishery exploited the salmon stock in 1985 (4 ± 2%) was less than in 1983 and 1984 (11 ± 3% in both years). This shows that the net and coble and rod and line fisheries combined removed a relatively small proportion (20%) of available fish. In other words, in 1983, for example, after the nets had taken their share, 90–96 fish remained of every 100 fish entering the R. Spey. After the rods had removed their share, 77–88 potential spawners remained.

In 1980–88, the anglers fishing the R. Spey took between 57 and 92% of all (net and rod) salmon caught before 1 May. One main reason for this is that spring fish are likely to have been available to the rods (but not the nets) for an extended period after 30 April. If it is thought that too few eggs are laid by spring fish, curtailing rod fishing might allow significantly more spring fish to survive.

The level of exploitation exerted by the net and coble fishery on the R. Spey is considerably less than on the R. North Esk. No comparable data are available for other UK rivers. Nevertheless, the production of salmon in the R. North Esk, measured on the basis of catch per kilometre of river has consistently been the highest of any east-coast river, and smolt production has been sustained.

An analysis of the data obtained from tagging experiments at the mouth of the R. Spey over the 3-year period 1983–5 showed that tagged fish remained in the netting zone for 3–5 days on average, even though the discharge levels in the summer, when most fish were tagged, ranged widely (20–65 m^3 s^{-1})

between years. The relatively low flow rates in the summer of 1983 and 1984 did not significantly increase the length of time during which fish remained in the netting zone. By contrast, the mean number of days (34–48) between tagging and recapture for rod-caught fish shows that fish were available much longer for exploitation by the rods than by the nets.

From an analysis of catch and flow data, the Salmon Advisory Committee found no evidence that fishing at times of very low flows in the Rivers Dee (Aberdeenshire), Don, Spey, Tay and Tweed enhanced the catches of salmon by either nets or rods. As a result, they found it unnecessary to consider what steps might be taken to control fishing at times of low flow (Anon 1990b, Smith 1990).

The levels of exploitation on several Norwegian river stocks have been much higher. For example, the rates of exploitation on the R. Imsa stock of 1SW and 2SW fish were 66–98% and 89–100% (Hansen 1988), and the rates of exploitation on both stocks (1SW and 2SW combined) in the Rivers Eira and Láerdaselve varied between 80 and 97% (Jensen 1979, 1981, Rosseland 1979). By contrast, low angling exploitation rates of 4% and 33% on the R. Drammenselv above and below a fish ladder respectively were probably an important cause of a rapid increase in the salmon population of this river (Hansen *et al*. 1986).

Mean exploitation rates by angling on four Icelandic rivers, the Ellidaar, Úlfarsá, Blandá and Nordurá, were 35%, 29%, 65% and 25% (Gudjónsson 1988). Levels of exploitation varied widely from year to year. On the R. Nordurá, for example, they ranged between 11 and 82% over a 13 year period. Exploitation rates by rod and line on the Burrishoole system, western Ireland, averaged 12% in 1970–81 (Mills *et al*. 1986), similar to that on the R. Spey.

On the R. Wye, the rod fishery took 25% in 1925–34 but 47% in 1965–74 when the number of licences increased from *c* 200 to 1300 Gee & Milner (1980). The exploitation level on the larger (>9 kg) fish may have been 100%. This suggests that the rate of exploitation on a declining stock may increase. Comparable data from other countries support this finding. Thus, the level of exploitation by rod and line on spring salmon in Scottish rivers may be considerably greater than the values quoted in Tables 9.1 and 9.2.

Because the level of exploitation on Scottish salmon stocks is generally so low, the detection of any effect which may result from changes in the management of the high seas or home water fisheries may require an unbroken time series of data extending in excess of 50 years.

9.3 The fate of North Esk salmon hatched in 1979

Research into the salmon population of the R. North Esk has for the first time reached a stage at which sufficient data are available to allow the fate of adult salmon derived from particular spawnings to be examined and quantified. Such a task requires unbroken data on smolt and adult salmon populations over a minimum period of 8 consecutive years. This can be difficult to obtain.

Data relevant to the 1978 spawning stock have been chosen to illustrate the method, primarily because these data are complete and typical. But equally importantly, the number of potential spawners in each of the years 1981–6 inclusive can be obtained directly and thus checked from the Logie counter records.

Information on the river and sea age composition, sex ratio and length distribution of the population of adult salmon entering the R. North Esk was obtained by sampling the net and coble catch on one day each week throughout the fishing season. The same biological characteristics of fish entering the river outwith the fishing season were obtained from sampling fish caught in Kinnaber Mill trap during the annual close time (1 September to 15 February). Simultaneous sampling of catches taken by these two methods had shown that the fish caught did not significantly differ from each other. Hence, the biological data from these two sources can be used to describe the biological characteristics of the potential spawning adult fish entering and ascending the river throughout the year. Total numbers are recorded by the automatic counter at Logie. Because such counter data were unavailable prior to 1981, the proportions of the 1978 1SW and MSW spawning stock assumed to have entered the river outside the fishing season were estimated from the relevant Logie counter and Kinnaber Mill trap data for 1981–7 to have been 47% for 1SW and 10% for MSW fish. These mean proportions have been applied to the 1978 data in order to obtain an estimate of the total number of potential spawners which entered the river in that year. Table 9.5 shows the estimated numbers of adult salmon, of each sea age group, which entered the R. North Esk in 1978, together with the losses to both the net and coble and rod and line fisheries. Table 9.5 also gives estimates of natural mortality in the river before spawning and non-catch fishing mortality. The latter mortality includes losses to poaching as well as the numbers of fish which were not caught but which were so damaged by the net or rod fisheries that they died before spawning. The last two categories cannot be quantified separately and they may vary considerably between years.

The number of potential female spawners surviving to spawn was estimated using the sex data obtained from the catch sampling programme. The relationship between length and fecundity determined by Shearer (1972) was then used to estimate the number of eggs deposited by each age group. Each female length was put into the fecundity equation to give the number of eggs produced by the females in the catch sample. Because it was assumed that the length distribution in the catch sample did not differ from that of the female spawners, the total egg deposition in 1978 (year S) was obtained by extrapolating this number of eggs to the estimated number of female spawners (Table 9.5). The length/fecundity relationship remained relatively stable in the late 1960s and early 1970s, because the group of fish whose eggs had been counted had all entered the river at about the same time and probably before June. However, recent data suggest that the fecundity of later running fish may be greater than that of fish of the same length which enter fresh water during the first half of the year, so that the figure for egg deposition may be an underestimate. From

Table 9.5 Estimated numbers of one, two and three sea-winter salmon entering the R. North Esk in 1978, the losses through fishing and natural mortality, the potential number of female spawners, and the estimated total egg deposition.

Sea age (winters)	Number of fish entering R. North Esk	Net and coble catch	Rod and line catch	Natural mortality in river	Potential number of female spawners	Estimated total egg deposition
1SW	15 352	3 615	136	173	5 463	17 174 369
2SW	7 735	3 558	882	173	1 967	10 825 006
3SW	337	155	38	0	91	822 064
Total	23 424	7 328	1 056	346	7 521	28 821 439

Table 9.6 Estimated numbers of smolts derived from the 1978 spawning in the R. North Esk.

River age (years)	Year of migration	Number of smolts produced
1	1980	3 036
2	1981	94 474
3	1982	44 960
4	1983	1 439
Total		143 909

the number of smolts estimated to have been produced in the R. North Esk in the relevant years (S + 2, S + 3, S + 4, and S + 5) and their age composition, the number of smolts derived from the eggs deposited in 1978 can be calculated (Table 9.6). These data suggest that of the 29 million eggs deposited, 0.5% survived to the smolt stage.

Fish from the 1978 spawning returned to the R. North Esk in 1981−1985 in proportion to the various combinations of river- and sea-age. Many of these fish carried tags which they had received as smolts. If it is assumed that the ratio of tagged to untagged salmon taken in all the fisheries which exploit salmon from the R. North Esk was the same, and also, that the tag reporting rates were as agreed by the North Atlantic Salmon Working Group of the International Council for the Exploration of the Sea (100% in the R. North Esk and Montrose Bay, 75% in the Faroes and 50% elsewhere, Anon (1986a)), then the contribution to each fishery can be calculated (Table 9.7). The use of a ratio to estimate the contribution made by fish of North Esk origin to these fisheries obviates the need to calculate both the number of fish which may have lost their tags and also when the loss occurred.

Of the adult salmon from the 1978 spawning which returned to Scottish home waters, 16 479 were caught in fisheries outside the R. North Esk. Of the

Table 9.7 Estimated catches of R. North Esk salmon in fisheries outside the R. North Esk in 1981−6.

Year	Faroes	W. Greenland	N.E. England and Ireland	Scotland	Estimated total catch
1981	0	0	0	111	111
1982	5	0	0	12 013	12 018
1983	597	0	497	2 229	3 323
1984	541	0	168	1 712	2 421
1985	73	71	0	414	558
1986	0	0	0	0	0
Total	1 216	71	665	16 479	18 431

Table 9.8 Fate of salmon which hatched in 1979 and returned to the R. North Esk in 1981−6.

Year	Number of salmon entering R. North Esk	Total catch	Number of potential spawners
1981	234	25	209
1982	6 107	1 672	4 435
1983	3 286	662	2 624
1984	4 245	688	3 557
1985	405	143	262
1986	23	9	14
Total	14 300	3 199	11 101

remainder, some may have spawned in other rivers (although the numbers involved are likely to have been extremely small), and some may have been lost to predation by seals but the rest entered the R. North Esk (Table 9.8). The latter number was obtained from in-river catch data and from the number of fish counted by the Logie fish counter. They were also described from the ongoing biological sampling programme. As a result, both the catches and the fish counted could be assigned to their respective age groups.

Of the North Esk salmon from the 1978 spawning which returned to Scottish home waters, 30 779 could be accounted for: 16 479 caught outside the R. North Esk and 14 300 entered the river. These returns are equivalent to a 21% survival rate between the smolt stage and return to home waters and 9.9% between smolting and return to the river (Fig. 9.10).

Salmon entering the R. North Esk are subject to losses by net and rod fisheries, from natural mortality and fishing mortality, as were the fish in their parents' generation. At present, only the catches and the escapement over Logie can be quantified. Of the 14 300 salmon which returned to the R. North Esk, 3199 were lost to fisheries. This leaves a potential spawning stock of 11 101 salmon. From these data, the survival rate between eggs and the spawners they produced was 0.04% and survival between smolts and spawners 7.7%.

9.4 Summary

In 1952−88 grilse (mainly) caught in bag nets at 13 sites around the Scottish coast were tagged before release. On the west coast there was little evidence of stocks from particular rivers making major contributions to a particular fishery while elsewhere fish from neighbouring rivers featured prominently in the list of recaptures.

The main direction of movement differed between tagging areas. Generally, fish move north on the west coast, east along the north coast, south on the

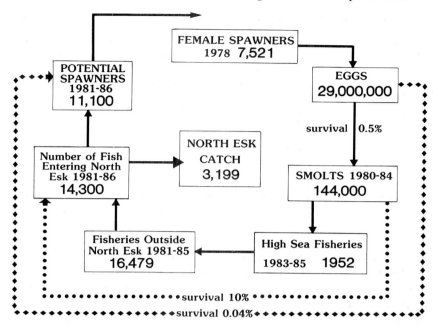

Fig. 9.10 The fate of R. North Esk salmon which hatched in 1979.

north-east coast, east along the south shore of the Moray Firth and north on the east coast. This pattern presupposes that, on the east coast, the fish must move south outside the range of the coastal fisheries before turning north again.

The levels of exploitation by net were lowest on the west coast (4% by fixed engine and 3% by net and coble), and highest on the east coast (29% by fixed engine and 26% by net and coble). These values were directly correlated with the densities of nets fishing. The length of time which fish spent in the netting zone was not affected by river flow.

Exploitation rates, on the fish available, by the net and coble fishery in the R. North Esk in 1976−88 were 30−63% and 35−62% for salmon and grilse respectively. However, because a varying proportion (>50% in some years) of the grilse spawning stock enters fresh water after the end of the fishing season, the level of exploitation on the total grilse stock entering the river was 15−40%. For several rivers, exploitation rates by anglers over the whole river averaged 5−8%. On the R. Spey in 1983, for example, nets removed 7% and anglers another 11% of the available fish. The exploitation rate by anglers in the R. Spey was low compared with some other rivers. However, the level of exploitation on the spring component of the stock may be considerably higher.

The levels of exploitation on several Norwegian and Icelandic river stocks were much higher. For example, rates of exploitation on the R. Imsa (Norway) stock of 1SW and 2SW fish were 66−98% and 89−100% respectively.

In 1978, 7521 spawners deposited 29 million eggs in the R. North Esk. From these eggs, a total of 144 000 fish (0.5%) survived to migrate to sea as one-, two-, three- or four-year-old smolts up to 1984. In 1983–85, the fisheries (west Greenland, Faroes, Ireland and Northumbria) removed 1952 fish outwith Scottish home waters. As the survivors made their way back to the R. North Esk in 1981–5, an additional 16 479 fish were caught by coastal fisheries in Scottish home waters. Of the remaining 14 300 fish which entered the R. North Esk in 1981 to 1986, the net and coble fishery caught 3199 leaving 11 100 potential spawners. These returns are equivalent to a 21% survival rate between the smolt stage and return to home waters and 9.9% between smolting and return to the river. The survival rate between eggs and the spawners they produced was 0.04%; between smolts and spawners it was 7.7%.

CHAPTER 10

CATCHES PAST AND PRESENT

Published catch statistics are only available for England and Wales and separately for Scotland since the early 1950s. However, historical catch records for particular rivers or fisheries, many of them dating back to the mid-1800s, have been deposited in archives or are held in private collections. Some of the owners of these private records have been extremely helpful in making their data readily available for analysis and a number of NRAs in England and Wales have done likewise. Mainly because of differences in the statutory requirements of the relevant Acts in Scotland and England and Wales, the data requested from fishermen and the manner in which the catch figures are summarized before publication are markedly different.

10.1 Scotland

10.1.1 The collection of present catch data

Before 1952, there was no statutory obligation to make data on catch and effort available to the Department of Agriculture and Fisheries for Scotland (DAFS). However, since 1952, catch data has been provided annually, in confidence, to the Department as a statutory requirement under the Salmon and Freshwater Fisheries (Protection) (Scotland) Act 1951 now amended by the Salmon Act 1986. Catches, divided between salmon, grilse and sea trout are recorded by the owner or lessee of each fishery by number, weight and method of capture (fixed engine, net and coble, or by rod and line). In addition, net and coble fisheries are requested, although this is not a statutory requirement, to supply the minimum and maximum number of netting crews and persons engaged in netting operations each month. Similarly, the operators of fixed engine fisheries are also asked for the minimum and maximum number of traps operated each month. Operators of fixed nets are also requested to give some additional details of gear used. For example bag, fly, jumper and stake net fishermen are asked to supply details of the maximum number of bags or pockets fished in any month. Poke net fishermen are asked for the maximum number of pokes, and haaf net fishermen for the total number of permits. Operators of more than one fishery in the same fishery district can combine their returns on one form.

151

Brief summaries have been published each year giving the reported catch for Scotland as a whole both between salmon, grilse and sea trout and between fixed engine, net and coble, and rod and line. In response to public demand, more detailed catch statistics were made available in 1981 covering the period from 1952. Thereafter, the detailed catch statistics were published on a regular basis. Although the 1951 Act provides that catch figures can be published for each fishery district, figures from some of the smaller districts, for reasons of confidentiality, have been amalgamated before publication to give a total of 62 statistical districts. These have been grouped into 11 geographic regions: East, North East, Moray Firth, North, North West, West, Clyde Coast, Solway, Outer Hebrides, Orkney and Shetland (Fig. 10.1).

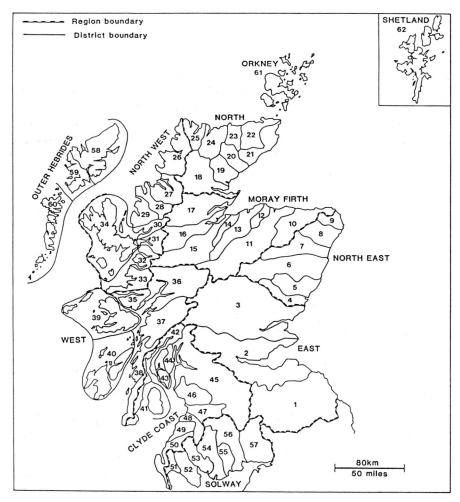

Fig. 10.1 Regions and districts defined for the analysis of catch statistics of salmon in Scotland.

Catches taken in each statistical district have been summarized by years (and for each year by statistical districts and regions), total weight and numbers of fish, and the all-method catch is given in separate figures for spring salmon (fish caught up to and including 30 April), summer salmon (fish caught on or after 1 May), and total salmon (spring plus summer), grilse, salmon plus grilse, and sea trout. Another set of tables shows the total catch in each region, first by all methods (i.e. total) and then by rod and line, net and coble and fixed engine fisheries. The final set of tables describes the 'all-Scotland' catch by year and by method, divided between spring, summer, total salmon, grilse, salmon plus grilse and sea trout. In the more recent statistical bulletins, there is an additional table giving the number of questionnaires sent to proprietors and occupiers of salmon fisheries and the number of returns received. The return rate in 1989 was over 95%.

10.1.2 *Present catches*

Catches are minimum values for two main reasons. First, they are incomplete, and secondly some fisheries, especially rod and line, under-report their catches.

Total catches of salmon in Scotland increased in 1962 and the new catch level of over 400 000 fish was maintained until about 1976. Since 1976, however, total catches have declined and recent figures are among the lowest since 1952. This decline was largely confined to net fisheries with little change in numbers caught by rod (Fig. 10.2).

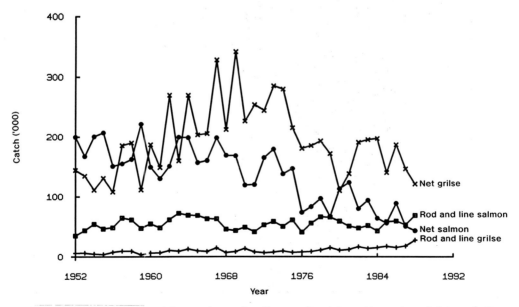

Fig. 10.2 Reported Scottish catches of grilse and salmon by net and by rod in 1952–88.

The reported Scottish catches from 1952−88 have been summarized in seven groups; the first 30 years in six groups (each containing five years) and the remaining seven years (1982−8) in the seventh group. Grouping the catches in this manner masks the wide fluctuations in annual catches but the groups show the general trend.

Catch trends differed between different components of the salmon stock. Whereas the total catch of spring fish dropped steadily from a mean catch of 88 000 in 1952−56 to a mean of 14 500 in 1982−88, the all-method catch of summer fish peaked in 1962−6 at a mean value of 182 000 before declining to a mean catch of 108 000 (34 000 less than the 1952−6 mean) in 1982−8. The trend in total numbers of grilse rose from a mean of 132 000 fish in 1952−6 to a peak mean value of 283 000 in 1967−71. It has since declined to a mean of 187 000 fish in 1982−8. The grilse mean catch was still 55 000 more than the comparable figures for 1952−6 (Fig. 10.3).

Rates of decline were not constant throughout these series of data. Whereas the decline in the five-year mean numbers of spring salmon in each period

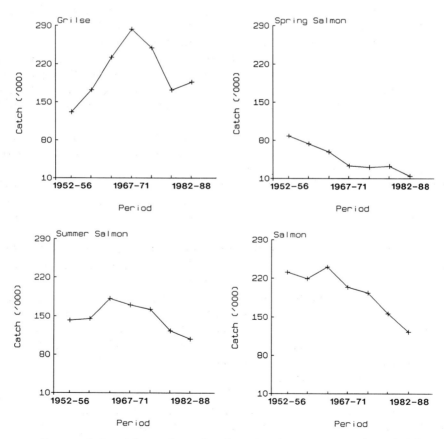

Fig. 10.3 Reported Scottish catches of grilse, spring, summer and total salmon in 1952−88, expressed as five-year means.

was marked and, apart from 1977−81, continuous (88 000−14 500), the decline in the numbers of summer fish was less marked (142 000−108 000) and did not commence until 1967−71 (Fig. 10.3).

In the catch returns submitted to DAFS, lessees and owners of salmon fisheries, with few exceptions have separated their catches into salmon and grilse on the basis of weight. Fish weighing less than 3.6 kg (8 lb) (in some fisheries 3.2 kg (7 lb)) have been classed as grilse and the rest as salmon when fish in both sea age groups were present. In general, those years when grilse were most abundant were also characterized by above-average proportions of over-sized grilse (Fig. 10.4). This error in classification enhanced summer salmon catches and depressed grilse catches.

In 1952−88, excepting 1987 and 1988, fixed engine and net and coble catches varied between about 110 000 and 280 000 in different years; the corresponding angling catch rarely exceeded 80 000. However, in 1987, due to a marked decline in net catches, the differences in the catches taken by rod and line (70 000) and net and coble (84 000) and fixed engine (114 000) were much less and in 1988, the rod catch (96 000) exceeded both the net and coble (80 000) and fixed engine catch (85 000) for the first time. The decline in catches was therefore borne by the net fisheries (Fig. 10.5).

In 1952−88, catches of grilse, spring salmon and summer salmon by net and by rod and line fluctuated independently. Mean grilse catches by net increased to a maximum in 1967−71, and then decreased. The pattern of angling catches of grilse was rather different. Except for a drop in 1972−6, catches of grilse by angling have steadily increased over the time series (Fig. 10.6).

Catches of spring salmon by net and by rod both declined in 1952−88, but the rate of decline was different in each case. Thus, net and rod catches fell by 92% and 47% respectively (Fig. 10.6).

Mean rod and line and net catches of summer salmon peaked in the 1960s and have subsequently declined. Although the rod catch in 1982−8 exceeded the corresponding value in 1952−6, the equivalent net catch was considerably less (Fig. 10.6).

Fixed engine catches contained the highest proportion of grilse (44−80%), and rod and line catches the lowest (11−25%). The proportions of grilse in the catches by all three gears increased in 1952−88.

In 1952−6, the proportions of the total Scottish catches of spring salmon, summer salmon, total salmon, grilse and salmon plus grilse combined taken by angling were 18%, 21%, 20%, 4% and 14% whereas in 1982−8 the corresponding values were 59%, 42%, 45%, 10% and 23%. Thus, the rod and line share for each of these groups increased over the period 1952−88, in one instance by more than three-fold.

There are a number of biases in the catch data from year to year which prevent their use as reliable indices of the strength of spawning stocks. First, salmon runs do not necessarily conform with fishing seasons. In 1981−8 (the years for which data are available), an increasing percentage (over 66% in 1988) of the fish which returned to the R. North Esk moved upstream over the

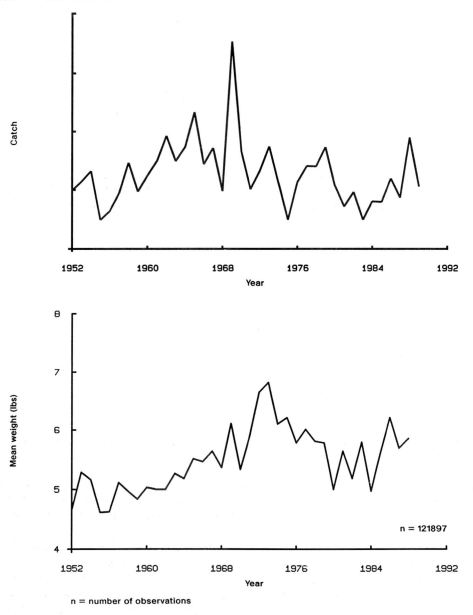

Fig. 10.4 Number and weight of grilse caught in the R. North Esk by net and coble in 1952−88.

Logie fish counter (5 km upstream from the sea) after the end of the net fishing season. Fish which arrive on the coast and enter rivers after the end of the netting season contribute neither to the fixed engine catch nor to the catches taken by net in rivers.

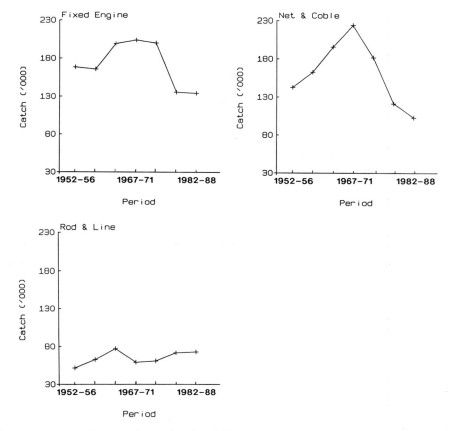

Fig. 10.5 Reported Scottish catches by different gears in 1952–88, expressed as five-year means.

Due to the lack of details on the variation in the effort expended in catching the fish, it is not known whether trends observed in the catches are correlated with the availability of fish or with the effort put into catching them or with neither.

The rod catch is likely to be sensitive to physical changes in the river system, such as the gravelling of pools, changes in the rates or temperatures of discharges and natural climatic fluctuations. In addition, any factor which alters the length of time which fish spend in fresh water, such as changes in the timing of runs within a fishing season, may affect the catchability of fish. Furthermore, the level of expertise of the angler can have a marked effect on the level of angling exploitation. Anecdotal evidence from ghillies suggests that whereas there have been great improvements in angling tackle in recent years, many of the anglers fishing our rivers are less expert than in former years.

Netting effort has declined, particularly in the most recent years (Table 10.1). Most fishing stations which are still active have shortened their fishing

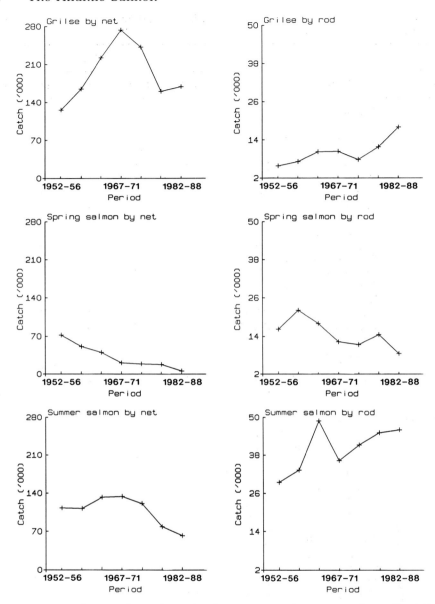

Fig. 10.6 Reported Scottish catches of grilse, spring and summer salmon by different gears in 1952−88, expressed as five-year means.

season, now commencing in late April or early May instead of February. On some rivers, the number of anglers fishing for salmon has increased, but on other rivers, the recent development of 'time sharing' of fishings has limited the number of anglers who can now fish these stretches. Even in the absence of 'time sharing' or 'syndication', angling effort on some rivers has peaked and

Table 10.1 Netting stations fished in the Dee, Bervie and North Esk Districts in 1955, 1965, 1975, 1985 and 1988.

Fishery district	Station		Year				
			1955	1965	1975	1985	1988
Dee	Foreshores						
	Girdleness and	(F)	*	*	*	*	—
	Greyhope						
	Bay or Nigg	(F)	*	*a	*a	—	—
	Altens	(F)	*	*	*	—	—
	Cove	(F)	*	*	*	*	*
	Cairnrobin	(F)	*	*b	*b	*b	*b
	Findon	(F)	*	—	—	—	—
	Portlethen	(F)	*	*	*	*	*
	Newtonhill	(F)	*	*	*	*c	*c
	Cowie	(F)	*	*	*	*	*
	Downie Point						
	Dunnottar and	(F)	*	*	*	*	*
	Gallaton						
	R. Dee						
	Midchingle	(NC)	*	*	*	*	—
	Pot and Fords	(NC)	*	*	—	—	—
Bervie	Catterline	(F)	*	*	—	—	—
	Shieldhill	(F)	*	—	*	—	—
	Bervie	(NC & F)	*	*	—	—	—
	Gourdon	(F)	*	—	—	—	—
North Esk	Johnshaven	(F)	*	*	*	*d	*d
	Rockhall	(F)	*	*	*	*	*
	Woodston Boat	(F)	*	*	*	*	*
	Woodston Fly	(F)	*	*	*	*	*
	Kirkside Boat	(F)	*	*	*	*	*
	Kirkside Fly	(F)	*	*	*	*	*
	Commieston	(F)	*	—	—	—	—
	Watermouth	(F)	*	*	*	*	—
	Charleton Boat	(F)	*	*	*	*	—
	Charleton Fly	(F)	*	*	*	*	*
	R. North Esk						
	Craigo	(NC)	*	*	—	—	—
	Morphie	(NC)	*	*	*	*	*
	Kinnaber	(NC)	*	*	*	*	*

F   fixed engine
NC net and coble
a   reduced fleet fished
b   reduced fleet fished
c   reduced fleet fished
d   reduced fleet fished

is now in decline. On the River Ness, for example, the number of day and weekday tickets sold in 1980−1990 has dropped from 728 and 105 to 384 (53%) and 39 (37%) respectively. It is sometimes not appreciated that many anglers are willing to pay large rents for the privilege of fishing by themselves so that there may be little scope in Scotland for any significant increase in the number of salmon anglers on the major salmon rivers.

The catch statistics described are based on information reported to DAFS by the owner or occupier of each fishery in response to an annual questionnaire. It is accepted that there is unlawful fishing in Scottish waters. The size of the catch is unknown and it is likely to vary both between regions and years. Because such catches are not included in any official return, the reported catch certainly underestimates the actual catch made by lawful fishermen (both nets and rods). Veitch (1989) suggested that the annual undeclared catches in the Tweed District amount to 5600 salmon and 3200 sea trout, and that in 1982, for example, an additional 29 500 salmon and 16 500 sea trout were caught illegally mainly offshore, declining to 3000 salmon and 7000 sea trout in 1988. For comparison, the Tweed District catches reported to DAFS were 28 983 salmon and 23 645 sea trout in 1982 and 18 232 salmon and 12 501 sea trout in 1988.

The decline in the total reported catch in 1952−88 does not signify a decrease in the strength of the overall spawning stock. Apart from the major decline in netting effort, changes in the sea age composition towards the younger age groups have decreased the proportion of the total stock returning in the first half of the fishing season. This is reflected in the decline of the spring catch by net and by rod coupled with the increase in grilse catches by both fishing methods and the increasing proportion of the total stock returning after the end of the fishing season.

10.1.3 *Past catches*

In most walks of life, it is usual to remember the good and forget the bad. Fishermen are no different. They tend to forget most past years in which catches were poor. Every year, they seem to remember, was a good year. It is an illuminating exercise to examine statistics of long-term catches, if only to place present day catches into some perspective. Suitable catch data back to the mid 1800s or earlier are available for most of the major in-river net and coble fisheries. However, when interpreting historical in-river catch data it must be remembered that subsequent to *c* 1830, salmon were exploited before they became available to the net and coble fisheries. This change in fishing practice may explain part of the general decline at that time in net and coble catches. Comparable data for rod catches are presently only available for one river, the R. Thurso. The data examined are the in-river net and coble catches taken in the Rivers Spey, Dee and Tweed (which together in the 1980s accounted for about 30% of the Scottish net and coble catch), and the R. Thurso angling catch.

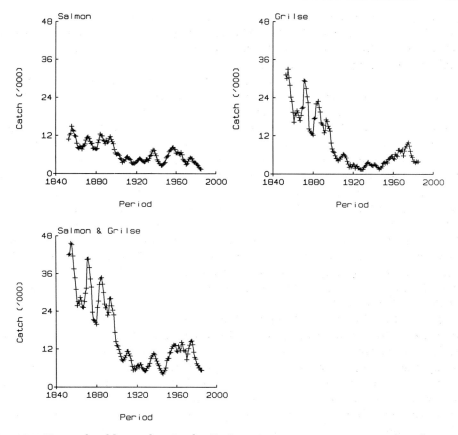

Fig. 10.7 Net and coble catches in the R. Spey in 1851−1987, expressed as five-year rolling averages.

10.1.3.1 River Spey

A comparison between the relatively recent (1952−87) and historical (1851−1951) data for net and coble catches shows that on the R. Spey, trends of grilse and salmon catches were dominated by changes in grilse numbers. Over the years 1851−1987 there was relatively little change in the catches of salmon (Fig. 10.7). There are limitations in these catch data, but analysis of the long-term trends reveals other interesting features. Changes occurred in the sea age composition of the catch; since the 1800s, the grilse component of the catch has first declined and then increased again. But the numbers of grilse caught have failed to reach the former high level of catches in 1851−70. Catches of salmon were higher in the 1800s than in this century, but the mean catch in 1900−39 was little different from that in 1940−87. Grilse catches have also declined since the 1800s; catches last century were about four times higher than in this century. Unlike the figures for salmon catches, however, the mean catch of grilse in 1900−39 was almost half the value for 1940−87 (Table 10.2).

Table 10.2 Catches of salmon and grilse by net and coble in the Rivers Spey, Tweed and Dee in 1850–1987, averaged for various time intervals.

Time Intervals	Mean catches					
	Salmon			Grilse		
	Spey	Tweed	Dee	Spey	Tweed	Dee
1850–99	10 010	11 205	1 068	19 928	15 281	897
1900–87	4 583	9 265	2 580	4 320	6 989	862
1900–39	4 715	9 248	2 435	3 524	4 877	634
1940–87	4 471	9 279	2 542	4 984	8 748	1 055

The seasonal catch distribution in 1987 was similar to that in the late 1800s except that the early spring catch in 1987 was a much smaller proportion of the total than formerly. Spring catches, mainly 2SW fish, tended to be above average when catches later in the season were below average (Fig. 10.8).

10.1.3.2 River Tweed

Analysis of the long-term (1808–1987) catch data for the R. Tweed net and coble fishery also reveals interesting features. Changes have occurred in the sea age composition of the catch. Following the large decline in grilse catches

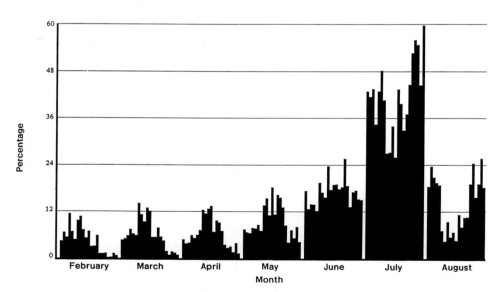

Fig. 10.8 Percentage of total net and coble catch in the R. Spey of salmon and grilse combined taken each month in 1887–1987, expressed as five-year rolling means.

in the late 1840s, total catches continued to decline at a much slower rate through the 1850s, 1860s and 1870s, both absolutely and as a proportion of the combined salmon and grilse catch. After 1880, a general increase occurred in the grilse catch. However, the higher grilse catches were not consistent, large and small catches sometimes alternating with one another on an annual basis. The improved grilse catches continued until the late 1880s before again declining until the 1920s. During the 1950s and for much of the 1960s, grilse catches increased and peaked in the late 1960s, but they have subsequently declined (Fig. 10.9).

On the other hand, salmon catches fluctuated more widely and more frequently. They peaked in the late 1810s, 1840s, 1880s, early 1900s, late 1920s, early 1930s and late 1950s and 1960s. But there was no evidence of a long-term underlying trend (Fig. 10.9). Data for the spring and summer components of catches were not available before 1900, and there is little evidence in the

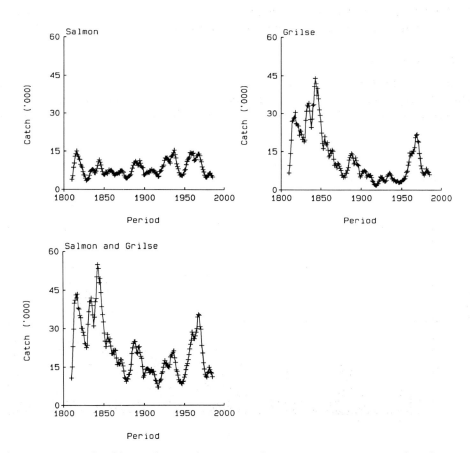

Fig. 10.9 Net and coble catches in the R. Tweed in 1808−1987, expressed as five-year rolling averages.

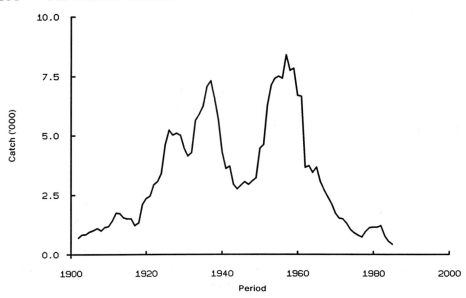

Fig. 10.10 Net and coble catches in the R. Tweed of spring fish in 1900−87, expressed as five-year rolling averages.

literature of a spring run of salmon from 1820 to 1900. The available evidence on the timing of salmon runs suggests that for much of the 1800s, the salmon run became later and later and that the last two weeks of September became proportionately by far the most productive weeks of the season. W. Henderson in 'My life as an Angler', published in 1880, reports that while fishing in the spring at Dryburgh, most of the fish he caught were generally from 3.6−6.7 kg (8−15 lb); all were kelts, and doubtless had been much heavier when they came from the sea in the previous autumn. The same angler, fishing on Rosebank (a beat much nearer the sea than Dryburgh) on 27 March 1849, caught 14 fish of which only three were fresh run. He adds: 'it is very unusual to take so many clean fish so early in the year'. In fact, there was a proposal at that time to delay the opening of the river above the estuary until March.

The importance of spring salmon to catches increased in the 1900s and by the late 1950s they were contributing 60% to the total R. Tweed salmon. However, a decline in spring catches has followed, and in the 1980s, before netting stopped in February and March, the contribution made by spring fish to the total salmon catch had declined to 18%. This is slightly above the corresponding value (15%) in 1900−10 (Figs 10.10 and 10.11). It was from this low base that spring salmon catches in the R. Tweed increased later this century. This increase augurs well for the future in showing that salmon catches can recover from relatively low to relatively high levels without significant assistance from man.

Although net fishing in the R. Tweed is limited to the first half of September, approximately 25% of the total annual catch in the 1980s was taken during

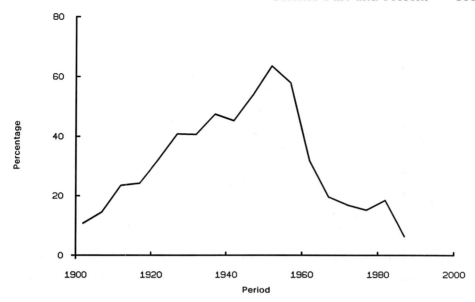

Fig. 10.11 Net and coble catches in the R. Tweed of spring fish in 1900–87, expressed as percentages of the corresponding total salmon catches.

this period compared with some 2% in the 1950s. The August and September catches together now account for more than 50% of the total annual catch (Fig. 10.12).

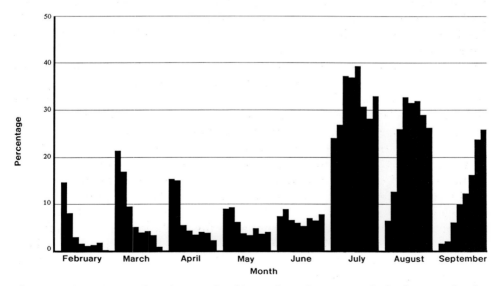

Fig. 10.12 Percentage of total net and coble catch in the R. Tweed of salmon and grilse combined taken each month in 1952–87, expressed as five-year means.

Catch patterns of grilse and salmon were similar in the Rivers Spey and Tweed in a number of respects. These included:

(1) An initial decline in the numbers of grilse and salmon caught by net and coble,
(2) The subsequent recovery period culminating in peak grilse and salmon catches in the 1960s,
(3) A subsequent decline in the catches of both sea age groups,
(4) Changes in the distribution of the catch between seasons.

Rather than a real increase in the numbers of MSW fish, the apparent peak observed in the late salmon catches from both rivers in the 1960s may have been the result of the inclusion of substantial numbers of overweight grilse in the salmon catches reported to DAFS. These fish were known to have been present in catches at that time.

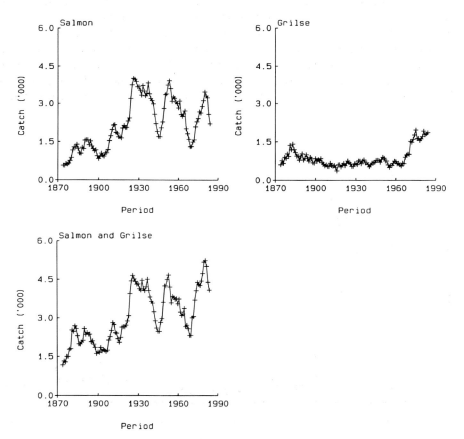

Fig. 10.13 Net and coble catches in the R. Dee in 1872−1986, expressed as five-year rolling averages.

10.1.3.3 River Dee

Trends in the numbers of salmon and grilse caught by the R. Dee net and coble fishery (Fig. 10.13) differed from those in the Rivers Spey and Tweed. Excepting declines in the late 1940s and early 1970s, the underlying trend in the R. Dee salmon catches in 1872–86 was upward. Catches peaked in the early 1920s, late 1930s, early 1950s and in 1980–84. The smallest mean catch since 1915–19 was in 1970–75, and the five-year mean catch in 1872–1986 increased from 560 in 1872–4 to 2210 in 1982–6. On the other hand, catches of grilse remained remarkably stable until 1960–64, since when they too have shown a steady increase (Fig. 10.13). However, low catches in the 1870s may have been caused by harbour works which were undertaken at that time. These diverted the estuary of the R. Dee, and may have prevented salmon from entering the river. It is not possible to analyse catches for numbers of salmon and grilse prior to 1872. However, the total weight of salmon landed each year between 1818 and 1986 indicates that catches steadily declined from the 1830s and reached a nadir in the 1870s (Fig. 10.14).

Two-year-old smolts formed the most numerous single age group in the catches in the 1920s and in 1983 (the only years for which data are available). However, approximately 78% of the catch in the 1920s had migrated to sea after spending two years in fresh water compared with 50% in 1983. The increase in the average smolt age was not limited to one particular sea age group but it was most marked in the 2SW age group (Shearer 1985b). Because fish derived from the older smolts tend to return in the first half of the fishing

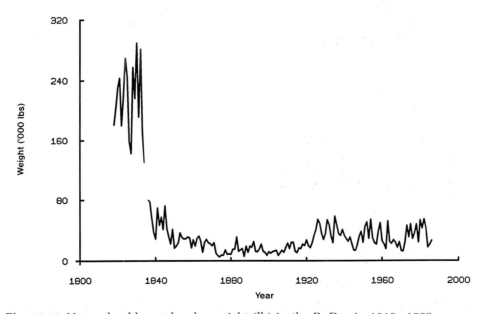

Fig. 10.14 Net and coble catches by weight (lb) in the R. Dee in 1818–1986.

season, this result suggests that a higher proportion of the catch was taken in the spring in 1983 compared with the 1920s.

The trend in numbers of grilse caught in the R. Dee also differed from those of the Rivers Spey and Tweed. Instead of a decline followed by an increase, grilse catches in the R. Dee changed little until the early 1960s, since when they have shown a steady increase. There was no decline in the 1980s. The increase in the five-year mean catch in 1872−1986 was approximately six-fold (552−3226) (Fig. 10.13).

Trends for the numbers of salmon plus grilse caught in the Rivers Spey and Tweed closely followed those for grilse. In contrast, the trend for the combined salmon plus grilse catch in the R. Dee closely followed that for salmon, and illustrates the much greater importance of MSW fish to R. Dee catches than to either the Rivers Spey or the Tweed (Figs 10.7, 10.9 and 10.13).

10.1.3.4 River Thurso

The trend of rod catches in the R. Thurso between 1881 and 1986 was upward. Mean catches peaked in 1961−5. Low catches in the war years, 1939−45, were the result of a combination of factors other than a big decrease in the number of available fish. Present catches, although less than those taken in the 1960s, are generally above the values recorded prior to the early 1920s (Fig. 10.15). However, in 1881−6, marked differences existed in the trends for numbers caught in January to April as compared with May to October. Whereas in January to April the trend was downward, the May to October catch trend was upward (Fig. 10.16). This pattern is similar to that for the rest of Scotland (Shearer 1986b). In the R. Thurso, not only was the catch

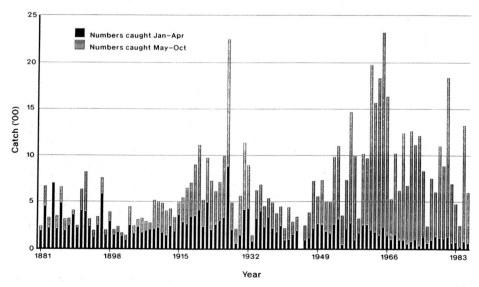

Fig. 10.15 Rod and line catches in the R. Thurso in 1881−1986.

recorded in 1923−86, but also the number of rods fishing each day including those who caught no fish. These data have been used to calculate the catch per unit effort (CPUE) which eliminates one of the problem variables when comparing catches between years (Fig. 10.16). The units chosen for comparison are the catch per 100 rod days.

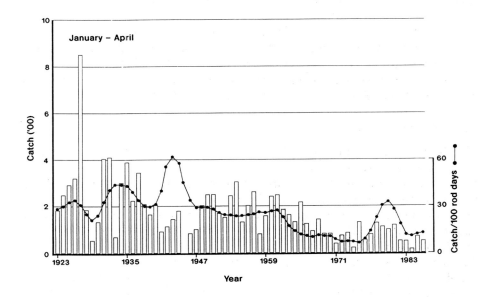

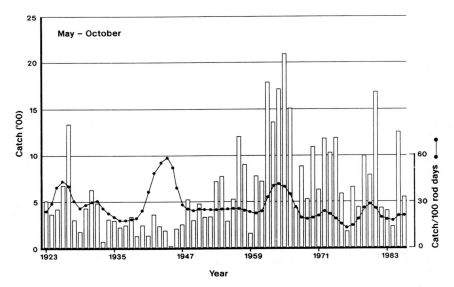

Fig. 10.16 Rod and line catches in the R. Thurso in January to April and May to October in 1923−86 together with catch per 100 rod days.

When the level of fishing effort does not change, catches and the corresponding CPUE follow the same general pattern. In these circumstances, the catch by itself may be a reliable index of the numbers of catchable salmon. However, this scenario seldom occurs. For instance, the temporary increase in CPUE in the 1940s showed that the major factor influencing catches in that period was a decline in fishing effort rather than a decrease in the number of available fish. On the other hand, above average catches in the period May to October, particularly in the mid-1960s, probably indicated a real increase in the number of fish available.

Although catches were not divided into grilse and salmon, the individual weights of all fish caught in 1887−1907 and 1923−86 (except for 1945 and 1974) have been recorded. For convenience these weights have been summarized into 3 groups: >3.6 kg (8 lb), 3.6−6.3 kg (8−14 lb), and <6.3 kg (14 lb). The proportions in each of these weight categories caught in January−April and May−October each year are shown in Fig. 10.17.

In 1887−1986, the main feature identified in the January−April period was a decline in the proportion of fish weighing more than 6.3 kg (14 lb). Although the proportion of fish weighing in excess of this amount also declined in the period May−October, this period was dominated, particularly from the 1940s onward, by a marked increase in the proportion of fish weighing less than 3.6 kg (8 lb).

Although it is impossible in the absence of an age/weight key to determine sea age from weight, the declines in both periods in the proportions of fish weighing >6.3 kg (14 lb), and the upsurge in the proportion of fish of this weight in the period May−October, were probably the results of a decline in the relative numbers of 3 or more SW fish and an increase in the numbers of 1SW fish. These suspected changes in the sea age composition of the catch were typical of the Scottish catch.

10.1.3.5 Historical Scottish catches

Records of the weight of salmon carried in 1894−1931 by the rail and steamship companies (Fishery Board for Scotland, 1932) illustrate the wide fluctuations in historical annual Scottish salmon catches. In the relatively short period 1894−1931, the weight of salmon and sea trout carried fluctuated between five-year means of 1650 t in 1914−18 and 2830 t in 1894−8, with no evident underlying trend. After adjustment for weight of ice and boxes and for double counting, these figures become 907 t and 1557 t respectively. The average weight of salmon (including sea trout) reported caught in 1983−88 was 1161 t, which lies between these two values.

10.1.3.6 Fishing seasons

When comparing long-term catch data, it should be remembered that fishing seasons have changed for both rods and nets. In 1852, for example, the season on the R. Dee lasted from 1 December to 20 September and in 1902 from 11 February to 26 August (nets) or 31 October (rods). More recently, the rod

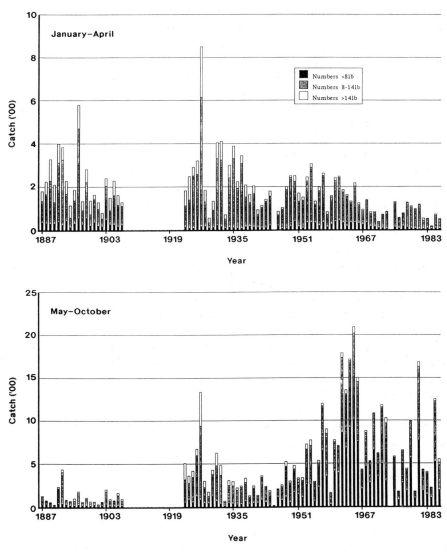

Fig. 10.17 Rod and line catches in the R. Thurso in January to April and May to October in 1887–1986, by weight groups.

fishing season was reduced by one month to 30 September. Formerly, the season on the R. Spey lasted from 11 February to 14 September, but from 1851 the netsmen, at first voluntarily and from 1862 by law, ceased fishing on 26 August while angling was permissible until 15 October. In 1948, the extension for angling was shortened to 30 September. The weekly close time for nets was increased by 6 h in 1952 and by an additional 18 h in 1988.

10.1.4 *The main features of the catch data*

Catches taken in the legal drift net fishery which operated off the Scottish coast in 1961 and 1962 are not included in the catch statistics submitted to DAFS. These national catches amounted to an estimated 28 000 fish in 1961 and 115 000 fish in 1962.

Scottish catches increased in 1962 and this new catch level was maintained until about 1976. Since 1976, total catches have declined and recent figures are among the lowest since 1952. The decline was confined to net fisheries. The numbers caught by rod more than doubled in 1952−88. Catch trends differed between different components of the salmon stock.

In addition to indicating wide fluctuations, the long-term catch data show that the mean catches of salmon in the Rivers Spey and Tweed were greater in the 1800s than in the 1900s, and that the mean catches did not change from 1900−1987. Mean grilse catches in the Rivers Spey and Tweed in the 1800s were more than twice those in the 1900s. Catch trends in the R. Dee were different. The mean salmon catch in 1900−86 was more than twice the corresponding value in the 1800s. But in the case of grilse catches, there was little difference between the mean values for the 1800s and 1900s (Table 10.1).

The lack of reliable data describing the effort put into catching salmon is a major difficulty in interpreting catch trends. The results of any comparisons between catches by years and by gears and their relationship to the spawning stock must be suspect. An added difficulty is that since the early 1960s, Scottish salmon stocks have been exploited outside Scottish home waters. The data from these fisheries are insufficient to assess losses to Scottish home water catches. The non-reporting of catches by legal and illegal methods has also fluctuated widely over the period.

Although angling catches remained markedly stable, the distribution of the catch, in common with the net catch, has shifted. In 1952−6, 32% of the mean annual angling catch was taken before 1 May but only 12% in 1982−8. An increase in the proportions of the angling catch taken after 30 April tends to mask the decline in spring catches.

10.2 England and Wales

10.2.1 *The collection of present catch data*

Although the Salmon and Freshwater Fisheries Act 1923 gave RWAs powers to make a bye-law requiring 'persons taking salmon and trout to make a return', it would appear that few people made use of this facility. As a result, national data for rod and line and net catches seem to have been based largely on the diaries maintained by the enforcement staff or on returns submitted by the private owners of individual fisheries on a number of rivers such as the

R. Wye. It was not until after the Salmon and Freshwater Fisheries Act 1972, which expanded the powers granted by the previous Act to make a bye-law requiring the submission of a return of catch 'in such form, giving such particulars and at such times as may be specified', that the direct submission of catch returns by the individual concerned (netsmen or angler) became mandatory in most regions.

In addition, the 1972 Act included a provision to require the submission of a 'nil' return of catch. The methods of collecting the catch data have significantly changed in the 39 years 1951−89. While the compilation of annual catches by each licensed instrument (rod or net) from the mandatory returns submitted directly to the NRA is now common practice throughout England and Wales, this type of approach is relatively recent and the level of commitment is not yet universal across all NRA areas. Furthermore, there are marked differences between Regions in the format and complexity of the catch return, the amount of detail requested and the issuing of formal reminder notices. For example, the introduction of a catch-reminder system for anglers in Wales in 1976 raised the total declared rod catch for England and Wales by some 10% in subsequent years (Harris 1988). This author also describes the results of a study undertaken in Wales in 1985 recording the number of salmon and sea trout observed to have been taken by selected fishermen. These observations were then compared with the actual return made by these fishermen. Some fishermen submitted an accurate return of their catches, but some others did not. The true catch was three to four times greater than the catch declared to the Authority.

In 1952−72, the number of licensed anglers increased by about 81% (Anon 1974). No more recent comparable data are available but the trend has probably continued. Angling effort has trebled since 1952. The level of increase has not been the same across the country. For example, on the Rivers Usk and Wye and within Wales the percentage increases in rod licence sales have been 100, 400 and 1400 respectively. In addition, individual anglers may fish more frequently than previously and few waters are now restricted to 'fly only'.

On the other hand, since 1952, the number of netsmen throughout England and Wales has remained relatively constant. Net Limitation Orders have been revived or introduced, but where changes have been made, they have been minimal and only of local significance. There has been one major exception: the dramatic increase in the drift net fishery in the 1950s and 1960s in Northumbria. In 1953, the catch declared for 58 licensed drift nets was 2006 salmon or 6% of the total catches in England and Wales. By 1970, there was almost a four-fold increase in the number of licensed drift nets (58 to 218) and a 45-fold increase (2006 to 90 587) in catch. As a result of these increases, this fishery was now catching 72% of the declared total commercial catch for England and Wales. The drift net fishery has now been limited to 121 licences and the licensee must normally be present when the licensed net is in operation. Nevertheless, in 1987, although the reported catch had declined by about one-third to 30 345 fish (44% of the England and Wales commercial catch), it was still 15 times higher than in 1952.

In the early 1950s, catch data for some rivers were not included in the total for England and Wales because the number of fish caught were not recorded. Catch data for 1951−82 are still regarded by the Ministry of Agriculture, Fisheries and Food as provisional. However, the overall record has become progressively more complete.

Illegal fishing is the biggest threat to the future status of the salmon fisheries of England and Wales (Anon 1983). Few would challenge the general statement that the illegal catch may now exceed the legal catch on many English and Welsh rivers. This significant change has occurred over the last 15 years with the extension of illegal fishing to estuaries and coast when the use of nylon mono-filament gill nets spread (Harris 1988).

Because catches are not divided into salmon and grilse, two methods have been employed for estimating the sea age composition. For most NRA areas, available catch data were divided into size categories. Scale sampling then provided an age/weight key. This key provided estimates of the relative proportions of grilse (1SW fish) and salmon (2+ SW) within each weight group. Where catch data were not stratified in weight categories, the age composition was estimated from the mean weights of grilse and salmon in the scale samples, according to the following formula:

$$N_g = \frac{N_p(w_s - W_p)}{w_s - w_g)}$$

where N_g = estimated weight of grilse in catch, N_p = declared catch of salmon and grilse, W_p = mean weight of declared catch, w_g = mean weight of grilse in samples, w_s = mean weight of salmon in samples, and: estimated number of salmon in catch (N_s) = $N_p - N_g$

This approach was used in preference to that of multiplying the catch by the proportions of grilse and salmon in the samples, because it is believed that the scale samples tended to be biased towards large fish (Russell & Buckley 1989).

10.2.2 *Present catches*

The catches taken by drift net within the area managed by the National Water Authority Northumbrian region fluctuated widely in 1965−87. This fishery exploits salmon returning to Scottish east coast rivers south of Stonehaven, principally the River Tweed. The underlying trend has been upward although less marked in the salmon component of the catch (Fig. 10.18). The more recent catches have shown some decline and the 1987 catch was 36% below the five-year average. A decline in fishing effort rather than any decrease in the numbers of catchable salmon may have been one of the contributory factors.

In common with Scottish rivers, the catch patterns for different English and Welsh rivers in 1951−88 have varied and the variation noted does not appear to be associated with the geographical location of the river. For example, total

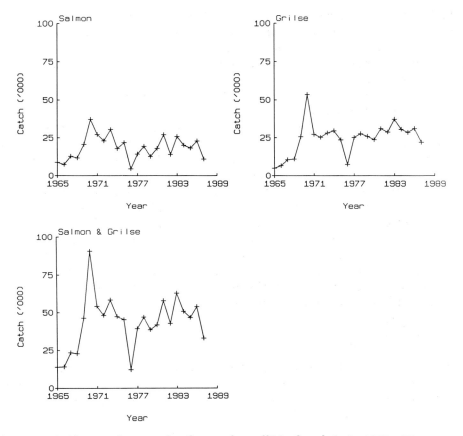

Fig. 10.18 Drift net salmon and grilse catches off Northumbria in 1965–87.

salmon catches in the R. Lune peaked in the mid-1960s and have since declined, while the R. Ribble catch data, although fluctuating more widely than that in the R. Lune, showed an underlying upward trend over the same period (Figs 10.19 and 10.20). Total catches in the R. Exe peaked in the 1950s and 1960s, declined throughout the second half of the 1960s and the first half of the 1970s, and then increased through the 1980s to a level which is little different from the peak catches in the mid-1960s (Fig. 10.21).

Catches by nets and rods have generally not followed the same pattern. Rod catches taken in the R. Tyne increased steadily since the mid-1960s and the rate of increase in the 1980s was exceptional (Fig. 10.22). For instance, the 1987 rod catch was more than double the five-year average. In the R. Ribble, after remaining relatively stable until 1970, rod catches increased and in the 1970s and 1980s the underlying trend was upward. By contrast, rod catches in the R. Exe peaked in the mid-1960s, declined throughout the remainder of the 1960s and early 1970s and then recovered in the 1980s. However, they have

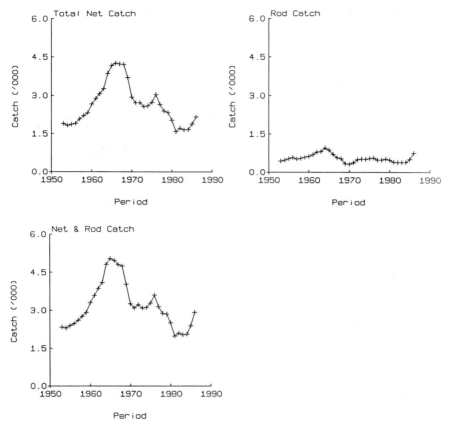

Fig. 10.19 Salmon catches in the R. Lune in 1951−88, expressed as five-year rolling averages.

not yet attained the catch levels of the 1950s and early 1960s (Figs 10.20 and 10.21). Rod catches in the R. Lune also peaked in the 1960s, but in 1951−88, they were much more stable than in the other two rivers and the more recent catches are higher than those taken in the 1950s (Fig. 10.19).

Catches are influenced by many factors other than stock abundance during the fishing season. The more important factors have already been described in relation to the Scottish fisheries. In general, however, catches in nearly all rivers show a trough around 1979 following above average catches in the late 1960s or early 1970s. The widespread declines in the mid-1970s reduced catch levels to their pre-1960 levels. Since the late 1970s, however, most catches have shown an upturn. In the R. Tyne, the increase in the rod catch over the period was exceptional but the magnitude of this increase was partly due to its low base in the 1950s. This increase was probably due to the steady improvement of water quality in the estuary. This resulted from continuing

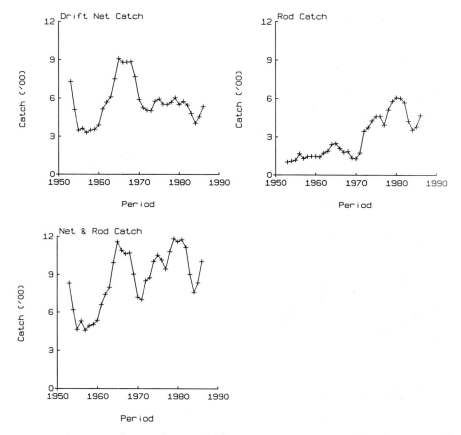

Fig. 10.20 Salmon catches in the R. Ribble in 1951−88, expressed as five-year rolling averages.

improvements in sewage treatment procedures, aided by the RWA enhancement programmes.

10.2.3 Past catches

Historical salmon catch records for the Rivers Avon, Stour, Frome and Piddle for the period until 1923 were compiled by the late J.D. Brayshaw and are published in the Bledisloe Report (Anon 1961). The data were drawn from individual estate fisheries and bailiffs. Since 1923, bye-laws passed under the Salmon and Freshwater Fisheries Acts made obligatory the reporting of annual catches in the area of the Avon and Dorset fishery boards and the present NRA region. It is not possible to ensure the accuracy of catch returns, but since 1950 over 95% of licences issued have been accounted for (Anon 1986c).

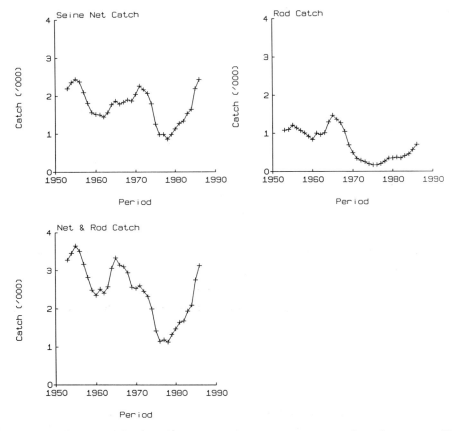

Fig. 10.21 Salmon catches in the R. Exe in 1951−88, expressed as five-year rolling averages.

Statistics from the mid 1800s show that catches fluctuated widely and often varied by ±50% in consecutive years (Fig. 10.23). Nevertheless, there has been no significant change in the 100−year trend for total catches in the Rivers Avon and Stour (Table 10.3). While catches were primarily by net in the 19th century, an increasing proportion has been taken by rods since the early 1900s.

In contrast, catches taken in the Rivers Frome and Piddle were low in the 1800s. This may have been due to the presence of a barrier weir trap at Wareham (Clark 1952). This weir was removed about 1900 and a noticeable increase in catches occurred thereafter (Fig. 10.24).

In addition to these long-term trends, short-term fluctuations in mean catches also occurred and peak catches in the Rivers Avon and Frome tended to coincide with one another. During the period since 1950, both rivers have shown a steady increase in yield to a peak around 1962, followed by an initial decline and a subsequent increase over the last 25 years.

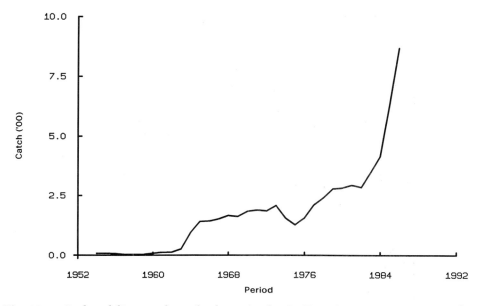

Fig. 10.22 Rod and line catches of salmon in the R. Tyne in 1952−88, expressed as five-year rolling averages.

In view of the changes in fishing effort, it is not possible to interpret these rod catches in terms of population trends. On the other hand, from the early 1900s onwards, when freshwater seining in the lower R. Avon ceased, net catch data came from a single fishery. They may reflect real changes in abundance during the fishing season.

The R. Wye is recognized as among the best salmon rivers in England and Wales, although it is reported that the stocks were greatly depleted in the 19th century by over fishing (Gee & Milner 1980).

Long-term catch statistics from 1905, including both rod and commercial catches, are available largely due to the enthusiasm of Mr J.A. Hutton. From 1910−40, total catches of both rod and net fisheries were generally similar. Subsequently, the net catch declined to average 1618 fish in 1975−7 whilst the rod catch increased (1975−7 mean 4691 fish), except in the more recent years (Fig. 10.25).

An investigation of the correlation between catches of 1SW, 2SW, and 3SW fish by nets and rods in 1909−1977 suggested that there was no immediate antagonistic effect between the two fisheries. In the long-term, catches of salmon by both methods may have been affected by a common factor, possibly stock abundance and environmental changes (Gee & Milner 1980).

The proportion of the older sea age groups, based on the number of fish in each weight group, has declined (Table 10.4). This change in the sea age composition of the catch had already been noted in catches from Scottish

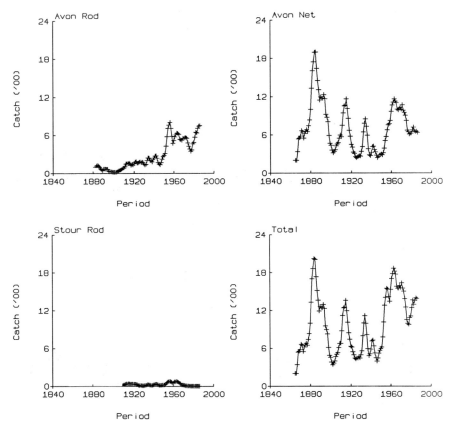

Fig. 10.23 Salmon catches in the Rivers Avon and Stour 1863−1988, expressed as five-year rolling averages.

fisheries. However, although the proportion of grilse in the catches in Scotland and the R. Wye was least in the period 1925−34, the R. Wye did not (unlike Scottish and Irish rivers) produce very large grilse catches later this century.

Table 10.3 Rivers Avon and Stour and Frome and Piddle salmon catch trends.

		100-year trend			Trend over last 15 years		
	100-year average Total	Slope and st. dev.	% change per year	't' Test	Slope and st. dev.	% change per year	't' Test
Avon & Stour	938	4.9(1.4)	+0.5	not sig. increase	−29(10)	−3.0	sig. decline
Frome & Piddle	287	5.9(0.6)	+2.0	sig. increase	−7 (2)	−4.0	sig. decline

Source: Anon (1986c)

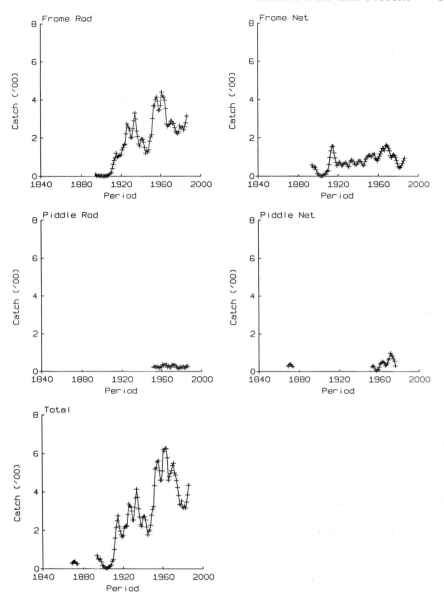

Fig. 10.24 Salmon catches in the Rivers Frome and Piddle 1867−88, expressed as five-year rolling averages.

10.2.4 *The main features of catch data*

In 1951−88, the trend of net catches in England and Wales was upward. They peaked in 1969 and have since declined, but present catch levels are higher than in the 1950s. Rod catches fell into two groups, 1952−66 and 1967−88.

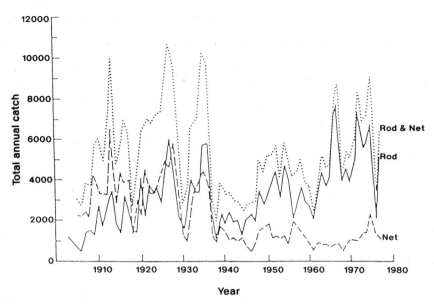

Fig. 10.25 Reported salmon catches in the R. Wye in 1905–77. (Reproduced from Gee and Milner 1980)

The catch in the first group was slightly higher than in the second group, but catches in each group were stable (Fig. 10.26). However, a critical analysis of the 38-year catch data for England and Wales leads to the general conclusion that these data provide neither a measure of the true catch nor a valid indicator of trends between years (Harris 1988).

10.3 Summary

Published catch statistics are only available for England and Wales and separately for Scotland since the early 1950s. Historical catch records for particular rivers or fisheries, many of them dating back to the 1800s, are held in private collections and some have been made available for study.

In the period 1952–88, reported grilse catches in Scotland peaked between 1967 and 1971. Although they have since declined, the mean catch in 1982–8 was still higher than the 1952–6 equivalent. Reported catches of MSW fish in 1982–8 were approximately 50% less than in 1952–6.

Within the MSW sea age group, the major decline occurred in the reported catches of spring fish. This sea age group dropped from a mean value of 88 000 fish in 1952 to 15 000 in 1982–8. The bulk of the decline was borne by the net fishery. In 1952–88, there was a two-fold increase in the proportions of the total Scottish catches of summer salmon, total salmon, grilse and

Table 10.4 Proportions of sea age groups (by weight) in R. Wye rod and net catches, five-year averages.

Year		1SW (%) (<3.2 kg)	2SW (%) (3.2−6.8 kg)	3SW (%) (6.8−13.6 kg)	4SW (%) (>13.6 kg)	Total (numbers)
1910−14	Rods	10.4	26.8	59.9	2.8	2373
	Nets	19.5	58.9	20.5	0.9	3831
1915−19	Rods	1.6	41.8	54.1	2.4	1796
	Nets	6.8	75.4	17.4	0.4	3422
1920−24	Rods	2.4	28.5	66.6	2.5	3193
	Nets	5.5	68.4	25.4	0.7	3334
1925−29	Rods	1.2	32.4	62.9	3.6	3449
	Nets	1.7	70.4	26.9	0.9	4811
1930−34	Rods	4.9	32.7	62.4	3.8	2620
	Nets	11.8	62.0	25.2	0.9	2349
1935−39	Rods	1.5	30.5	64.6	3.3	3329
	Nets	5.7	63.1	30.1	1.1	2262
1940−44	Rods	4.3	32.0	61.2	2.5	1706
	Nets	11.5	63.5	24.5	0.6	1026
1945−49	Rods	3.1	44.9	50.7	1.3	2436
	Nets	8.5	72.7	18.6	0.3	969
1950−54	Rods	3.5	45.2	50.2	1.0	3934
	Nets	12.6	74.2	13.1	0.1	1344
1955−59	Rods	5.6	49.1	44.6	0.7	3021
	Nets	11.4	78.5	9.9	0.2	1327
1960−64	Rods	5.4	50.1	44.0	0.6	3171
	Nets	21.6	69.6	8.7	0.1	628
1965−69	Rods	7.4	44.9	47.1	0.7	5441
	Nets	31.4	60.3	8.3	0	637
1970−74	Rods	13.1	56.4	30.0	0.6	5541
	Nets	35.0	59.3	5.6	0.1	1188
1975−77	Rods	12.6	63.5	23.5	0.5	4691
	Nets	32.2	64.8	3.0	0	1618

Source: Gee and Milner (1980)

salmon plus grilse combined taken by angling, and a three-fold increase in the proportion of spring fish. In 1988, the reported rod catch exceeded the net and coble and fixed engine catches for the first time.

Catch data also showed that in 1952−88 there was a general decline in the mean sea age of the catch and that the proportion of the annual total catch taken after April increased. Comparable changes had occurred in the 19th century.

In 1951−88, the trend of net catches in England and Wales was upward. This trend was influenced by the catch taken in the north-east coast drift net fishery because in most years this fishery accounted for more than 50% of the total net catch. Rod catches fell into two groups, 1952−66 and 1967−88. Catches in each group were markedly stable with the catch in the first group at a slightly higher level than in the second group.

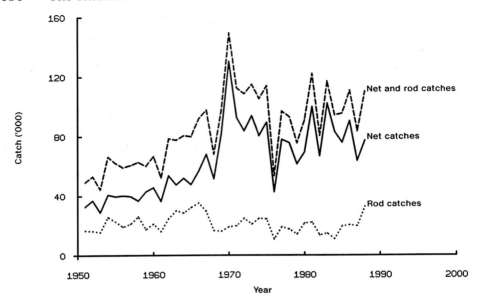

Fig. 10.26 Reported English and Welsh salmon catches by net and by rod in 1951−88.

Catch and stock are not synonymous and a number of biases in the catch data prevent their use as reliable indices of the strength of spawning stocks. These biases include changes in the timing of salmon runs so that they do not necessarily conform with fishing seasons, changes in the amount of fishing effort, changes in the level of exploitation and the catchability of fish both within and between seasons and from one area to another. Furthermore, the lack of reliable data describing the effort put into catching salmon makes suspect any comparisons between catches by gears and by years and their relationship to the spawning stock.

Historical data show that on the Rivers Spey and Tweed, trends were dominated by changes in grilse numbers. On the R. Dee, the trend followed that of MSW fish. On the R. Thurso, the trend of rod catches was upward between 1881 and 1987, due entirely to an increase in the catch taken from May to October. Historical catch data also showed that Scottish catches fluctuated widely over relatively short periods and that low catches were frequently followed by above average catches.

There is little evidence of a major spring run of salmon in the 1800s. Spring fish probably made their largest contribution to Scottish salmon catches in the 1950s.

The two factors which had most bearing on the level of recent total Scottish salmon catches have been the increased proportion of fish entering rivers after the end of the fishing season, and a decrease in net fishing effort. There is no evidence that Scottish salmon stocks are in serious decline or that egg deposition has limited smolt production where fish have had free access. In common with Scottish rivers, historical catch patterns for different English and Welsh rivers have varied. Catches by nets and rods have generally not followed the same pattern.

Nevertheless, taken at face value, the pattern illustrated by the English and Welsh net catches differs in a number of respects from that shown by the corresponding Scottish catch. The main difference is the underlying upward trend in English and Welsh catches since 1951 and the higher level of catches in the 1980s compared with the 1950s. Marked stability is the main feature of the rod catches in England and Wales and Scotland, albeit at two different levels in the case of the English and Welsh catches.

CHAPTER 11

SALMON FARMING

Commercial farming of Atlantic salmon in large sea enclosures was successfully pioneered in Norway in the mid-1960s by A/S Mowie. This success was followed a few years later by the Norwegian Grontveld Brothers on the island of Hitra using the now extensively adopted floating net cage system. Development in Scotland followed very quickly and the domestication of native Scottish stock was achieved by Marine Harvest at Lochailort (Inverness-shire). Commercial production began by the mid-1970s. In 1979, 500 t were produced from 14 sea water sites (Plate 19 in the colour section).

The main area for early developments was in the south west Highlands, mainly because of locational advantages. Sites were close to major population centres for supplies and marketing. Activities expanded into the north-west Highlands in the late 1970s. In the 1980s, expansion continued into the Western and Northern Isles. Restrictions imposed on the operations of big companies in Norway have diverted investment to Scotland. Recently, a number of Norwegian companies have invested in the Highlands and Islands, including the buy-out of one of the biggest Scottish companies, Golden Sea Produce, by Norsk Hydro.

The industry has grown to become a significant part of the Scottish economy, particularly in rural and isolated areas. The annual tonnages of salmon produced since records were first kept in 1979 show a considerable increase over the period (Fig. 11.1). In 1989, a total of 28 553 t was produced from 176 companies operating 331 sea-water sites. This tonnage is predicted to increase to almost 37 000 t in 1990. In addition, some 26 million smolts were produced by 90 companies operating 188 freshwater sites in 1989 (Figs 11.1 and 11.2). This number exceeds estimates of the average annual production of wild salmon smolts in Scotland. As a consequence, native salmon stocks, for the first time in their history, are presently greatly outnumbered by salmon of cultured origin.

This rapid growth has been paralleled by an increase in staff employed, from 144 in 1979 to 1418 in 1989, although there was a drop of 365 during the last year of that sequence. These figures include only those directly involved at the production sites. First sale values were estimated at £2 329 600 in 1979 and at £72 000 000 in 1988.

Salmon farming looks set to reach a plateau in production by 1991 through management decisions probably determined by market forces. However, even holding current levels of ova and smolt production, greater yields than the estimate should be achievable if sea water survival can be increased.

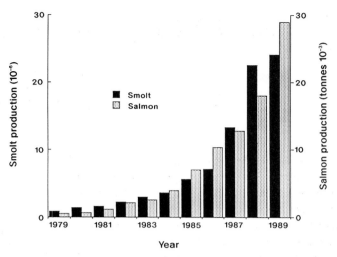

Fig. 11.1 Annual production of farmed Atlantic salmon and smolts in Scotland in 1979−89.

Salmon fisheries have been affected by the upsurge in salmon farming. The output of the Scottish commercial fisheries in recent years was *c* 1000 t per annum; farmed salmon output overtook this in 1981 and is now greatly in excess. An inevitable consequence has been the depression of the market price for the commercial fishermen during the second half of the fishing season when most of their catch is taken.

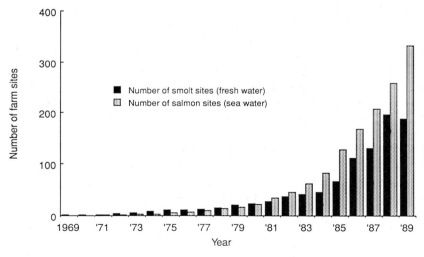

Fig. 11.2 Number of Atlantic salmon farming sites in operation in Scotland in 1969−89.

Salmon are believed to maintain distinct breeding populations with particular adaptive characteristics, even within major river systems. This is an important consideration for management of rivers on the west and north coasts close to fish farming enterprises. A large number of fish escaping from a fish farm mating with native stock may dilute the native gene pool and result in reduced viability of the 'hybrid' offspring. Escapees from fish farms are now regularly caught in coastal fisheries on the west coast and on some days escaped fish number over 10% of the catch. In addition, anglers now catch increasing numbers of escapees and such incidents are no longer limited to west-coast rivers. The visual inspection of catches underestimates the number of escapees present and this method misses particularly the fish which escape soon after they have been released into a sea cage.

Farmed fish may interfere with reproduction. For example, they may breed later and superimpose their redds on those of wild fish, or they may compete with wild fish for a limited number of suitable spawning sites. This might both reduce numbers of wild fish in the next generation as well as select for changes in spawning time. Ecological interactions will also occur between the offspring of farmed fish and those of wild fish, for example, competition for territories and food.

Unfortunately, few studies are available to shed light on the actual and potential impact of ecological interactions. Studies which have been carried out on interbreeding between salmonid stocks suggest that hybrids between native and hatchery fish will do less well than native fish in the wild. For example, juvenile survival and the probability of returning as an adult to the natal river to spawn are both less.

In autumn 1989, many salmon which had escaped from cages in Loch Eriboll (north-west Scotland) were observed spawning in the R. Polla. This river drains into the head of the loch and probably has a relatively small population of wild fish. Most escapees were seen spawning in the lowermost reaches and some had mated with native fish. At a fixed engine station near Gairloch in 1990, it was estimated that between 10 and 45% of each day's catch comprised fish farm escapees (Webb, pers comm).

In Norway, more than half the spawning stock in some major rivers are estimated to be fish farm escapees. These fish have been observed mating both with wild and farmed fish. As a consequence, the smolt production which is normally set by the rearing capacity of the river and not by egg deposition, may now contain a high proportion of 'hybrids' which will be less well adapted and whose survival in the sea is likely to be much reduced compared with native smolts. Thus, the number of adults returning from each 1000 smolts emigrating may be considerably less than previously. The productivity of the population will continue to decrease until selection can restore an optimum genetic constitution for the population. This may take generations. A gene bank for the preservation of the genetic diversity of threatened wild stocks has been established in Norway to counter threats from escapes from fish farms and losses due to parasitic infections and water acidification.

Conservative advice for those engaged in the enhancement of depleted salmon populations places importance on the effective use of the progeny of any remaining indigenous fish. This approach is most likely to conserve the genetic character of any stock and its intrinsic variability, ensuring its continued stability over long periods of time. Therefore, no foreign stock should be released in any river which still contains native fish.

Escapees can also occur in fresh water because a growing proportion of the smolts required for salt water cages are now being reared in freshwater lochs. With the suitable sites in west coast lochs full, salmon farming has spread to a number of lochs associated with major east-coast salmon rivers. Although considerable concern has been shown about the possibility of introduction or transfer of disease, there is no record of wild fish being affected by farm stock in Scotland (Munro *et al.* 1976, Munro & Waddell 1987). Nevertheless, Egidius (1987) records that the appearance of furunculosis in Norway coincided with the importation of Scottish salmon smolts. Furthermore, the skin parasite *Gyrodactylus salaris* was probably introduced to Norwegian salmon farms from resistant stock from Sweden (Johnels 1984). Subsequently it spread to wild salmon stocks in Norwegian rivers where 33 rivers are now affected. However, the view is expressed that it is a combination of introduced disease and the introgression of genes from cultured fish which has caused the marked reduction in salmon abundance in some rivers rather than disease alone.

Juveniles which do not smoltify at one year are superfluous to the requirements of most fish farmers. They are frequently accepted by district fishery boards to enhance the wild stocks in the rivers for which they are responsible either free or at a knock-down price. Except in the case of rivers in which stocks have either been lost or reduced to a very low level, the advantages of introducing foreign stock are doubtful. Ritter (1975) showed that the survival rates of hatchery-reared Atlantic salmon smolts released in various rivers were lower than those of native stocks. Altukhov (1981) attributed a marked decline in numbers of chum salmon (*Oncorhynchus keta Walbaum*) in the R. Naiba to 'disturbance of genetic structure' of the native population. This was supposed to have resulted from the introduction of eggs for several years from the neighbouring R. Kalininka. Before the transfer, the R. Naiba carried a spawning population of about 650 000 chum salmon. By 1969–70, the genetic characteristics of the stock returning to the R. Naiba had shifted towards those of the R. Kalininka fish. By 1980, the returning population had decreased to 30 000 to 40 000 spawners, and by 1985, the population was virtually extinct (Thorpe 1988). Altukhov (1981) concluded that this catastrophe was the result of massive genetic replacement by non-adapted genotypes.

The timing of the spawning migration in the Baltic Sea differs between wild and hatchery salmon. Salmon of wild origin migrate earlier than those reared in a hatchery (Ikonen 1987). Differences in the timing of the smolt migration in the R. Tummel between wild fish and fish planted out as unfed fry have been recorded by Struthers & Stewart (1986). Once again, the wild fish migrated earlier. In this instance, the timing was such that there was little overlap and the two runs were distinct. As a result, these introduced hatchery fish could influence the age composition of the adult stocks in the river into which they were released. This in turn could effect the availability of fish to the fisheries (net and rod) which must operate within a fixed season.

Cage rearing in inland lochs can add sufficient nutrient to the lochs to change their status from oligotrophic to eutrophic. As a consequence, the present fauna and flora may be replaced by animals and plants more suited to a richer environment. The growth rate of charr in Scottish lochs has been observed to increase suddenly after the rearing of juvenile salmon in cages commenced.

The presence of caged salmon in both freshwater and sea lochs may also attract predators which may take not only the caged fish but also the wild fish. The additional food supply may increase the survival of predators. On the other hand, predators attracted to fish farms may be short-lived because of shooting and accidental capture in the netting of the cages or they may be discouraged by the netting fixed outside the cages to protect their contents from predators such as seals.

11.1 Summary

Native salmon stocks, for the first time in their history, are presently greatly outnumbered by salmon of cultured origin as a result of the successful commercial farming of Atlantic salmon in large sea enclosures, successfully pioneered in Norway in the mid-1960s.

The development of fish farming on the Scottish west coast followed very quickly. In 1979, 500 t were produced from 14 sea water sites. Production increased very rapidly and by 1989, 28 553 t of salmon were produced artificially by 176 companies operating 331 sea-water sites. This tonnage was in addition to the production of 26 million smolts by 90 companies operating 188 freshwater sites. The projected production of farmed salmon in 1990 is 37 000 t.

The industry employs about 2000 people. It is now a significant part of the Scottish economy, particularly in the rural communities on the west coast and on the Western and Northern Isles. In 1988, the turnover was estimated at £72 million. An inevitable consequence has been to depress the market price of salmon for commercial fishermen. As a result, a number of netting stations have become no longer economically viable. These stations have closed down. The increase in local employment from salmon farming has not been all gain.

Salmon farming may reach a plateau in 1991 through management decisions probably dictated by market forces.

Farming has a number of important implications for river management, particularly if salmon maintain breeding populations with particular characteristics, even within major river systems.

If large numbers of escapees from a fish farm mate with native stock, the native gene pool may be diluted. This may result in reduced viability of the offspring. Recently, escapees have been observed in a Scottish river spawning in the wild both with other escapees and with wild salmon. There is also firm evidence, particularly from Norway, that escapees do migrate, do enter fresh water and do mate with wild fish. Farmed fish may enter rivers later than wild fish. As a result, farmed fish may overcut the redds made by wild fish.

At a fixed engine station near Gairloch in 1990 it was estimated that between 10 and 45% of each day's catch comprised fish farm escapees.

Escapages of juveniles from rearing cages in freshwater lochs, particularly those in the systems of major salmon rivers, give rise to the possibility of the introduction or transfer of disease. At present there are no records of wild fish being affected in Scotland. Nevertheless, the occurrence of furunculosis in Norway coincided with the importation of Scottish salmon smolts and the parasite *Gyrodactylus salaris* was probably introduced to wild salmon stocks in Norway from Norwegian fish farms rearing fish imported from Sweden. The transfer of salmon should be prohibited by law if necessary. Norway has established a gene bank to counter threats to the genetic diversity of wild stocks of salmon.

Except in the case of rivers in which stocks have either been lost or reduced to a very low level, the advantages of introducing foreign stock are considered doubtful.

It should always be assumed that any introduction of foreign stock is a real risk to the native salmon until it is proven that no such risk exists.

CHAPTER 12
CONFLICT WITH THE ENVIRONMENT

12.1 Changes in land use

In most rivers, the numbers of smolts produced is correlated with the quality of the nursery areas. The higher the quality of the nursery areas, the more smolts will be produced per unit area (Solomon 1985). The quality of nursery areas can be greatly affected by changes in land use.

12.1.1 Forestry

In the R. Spey catchment, for example, Rice (1988) identified a number of important changes in forestry practices which he suggested were amongst the most important to the appearance of the countryside during the 19th and 20th centuries. These changes include the following:

(1) Clearance of broadleaved woodland and Caledonian pine; large scale forestry planting in many parts of the middle and lower Speyside; increased appearance of silt in drainage systems,
(2) Lack of birch woodland regeneration arising from overgrazing,
(3) A steady improvement of in-bye land in the middle and upper parts of the valley and a significant improvement to the productivity of the farms in the lower part of the valley and coastal plain.

Often, a period of rapid change has been followed by a longer period of consolidation and while change was taking place in one form of land use, there may have been comparative stability in another.

Prior to the clearance of the climax woodland vegetation, Scotland was covered with trees — the bare mountains are 'man-made'. Nevertheless, by 1812 the area of woodland in Scotland was 37 000 ha and by March 1987, the productive woodland area in the catchment of the R. Spey alone amounted to approximately 33 600 ha (Calder & Gill 1988). The increase in acreage throughout this period was not at a steady rate. In 1850–91, the planting of 200 million trees by Seafield Estates was not sufficient to maintain pace with the rate of felling. The two world wars of the 20th century saw the decimation of the maturing forests of Speyside. However, the era between the wars and, more particularly, the period following the cessation of hostilities in 1945 saw the replanting of many deforested sites. Although the rate of new Forestry Commission planting has declined from the peak values in 1951–60, their total owned woodlands have increased dramatically from 1921 to 1986 (Fig. 12.1). Over half the plantation area originated in the period 1951–70, coinciding with the period when state afforestation reached its peak nationally. These values do not include private woodlands, and so the total amount planted has been considerably higher. Since about 1970, 20–44 ha have been clearfelled and replanted annually. Over the next ten years, this area will probably be halved until the areas planted in the 1930s reach maturity. After that, there should be a significant increase in the areas to be felled.

Afforestation, although perhaps not on the same grand scale, has occurred in other watersheds and, in common with the R. Spey catchment, many of the most favoured sites are located in the upper reaches.

In the post-war period, there were major advances in afforestation technology and changes in the relative proportions of Scots pine (*Pinus sylvestris*), Sitka spruce (*Picea sitchensis*) and Lodgepole pine (*Pinus contorta*) planted annually. These changes mostly paralleled the quality of land being planted.

McEwen (1985) discussed both the hydrological and sediment yield implications of such drastic land use changes as afforestation and deforestation. She considered the implications of land drainage on the varying volumes and frequencies of discharge, and presented evidence of increased sedimentation. Drainage associated with the double throw ploughs used by the Forestry

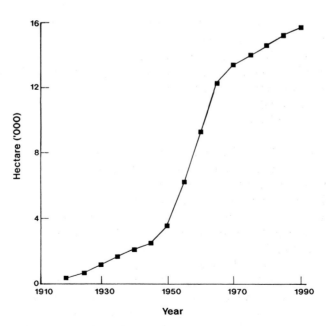

Fig. 12.1 Accumulated area (ha) of new Forestry Commission plantations in the R. Spey valley in 1921–86. (Reproduced from Calder & Gill 1988)

Commission can cause the rapid runoff of rain falling into the furrows. As 20% of the land surface in the Spey catchment may consist of ditches, 20% of any precipitation will fall directly into them and may be lost from the site very rapidly. The main effect is registered in the period immediately after ploughing. Peak discharges in the streams draining these areas have increased by 40% and the duration times of peaks have halved.

Position within the catchment, whether nearer the upper or lower limits, is clearly important. As a rule of thumb measure, it has been suggested that for every 10% of a catchment area afforested, the potential water yield drops by 2%. Losses will be greater in wetter, high-lying areas. Losses can be important locally where abstraction is from a small catchment with little or no storage and where summer base-flows can be crucial. It is generally accepted that burns of 1 m or less in width draining forestry plantations containing 20- to 30-year-old trees, will not have sufficient flow to support a salmonid population for 10 months each year.

The increase in peak discharge levels and in their frequency can cause erosion not only of the soil covering the surrounding land but also of the banks of the streams, thereby increasing their width. In addition, erosion can increase turbidity which adversely affects aquatic ecosystems in a number of ways. For example, suspended solids reduce light penetration and this may affect fish feeding and migration. Furthermore, the presence of suspended solids may directly affect fish respiration. When sediment settles in streams it makes watercourses more shallow, damages spawning areas, and modifies the substrate so that it can no longer support the invertebrates which form the major item in the diet of juvenile salmon. The gravel washed out as the banks erode settles in the holding pools in the main stem of the river to such an extent that a number of these pools may disappear together with their angling potential (Plate 20).

Changes in flow patterns may also have a marked effect on the timing of runs into rivers and the subsequent upstream migration. Entry into spawning burns may be delayed or physically impossible because of depressed flows. As a result, fish will spawn in the main river in an environment very different from that in the spawning burn. This change can alter the subsequent life history of the fish.

Chemicals, including herbicides, pesticides and fertilizers may enter the watercourses either directly or by leaching and this may adversely affect some invertebrates and the nutrient status of the stream. An association between afforestation and increased acidity of water has been observed in some areas overlying acid rocks and soils. This, together with reduced calcium and raised levels of aluminium, can result in the loss of invertebrates and fish.

The various deleterious effects on salmonid production can largely be minimized by careful forestry management. This fact has been recognized and has led to the publication by the Forestry Commission in 1989 of 'Forests and Water Guidelines'. In 1991 it was revised and a new section on timing was included.

This code of practice must be followed by forest managers. However, it

Plate 20 Destruction of a spawning and nursery area by silt deposition.

cannot resolve the effects of afforestation on flow characteristics. On the one hand, flows increase following initial drainage operations and on the other, the deeper root system of trees and their increased transpiration capacity result in lower levels during low flows. For example, in the valley of the R. Tweed, it has been shown that in adjacent catchments, the dry weather flow in an afforested catchment is only 50% of those in a grassland catchment (Fox 1989). If this is extrapolated in the same catchment using the projected increase in afforested area, the reduction in flows in the R. Tweed is likely to be somewhat greater than that due to reservoir development for public water supplies.

At present, forestry is the most rapidly expanding land-based activity in upland Scotland, and is likely to remain so in the foreseeable future. In addition, tree planting may spread to the less productive arable land presently producing mainly cereals. The conflict of interests, therefore, between the forester and the river manager is set to continue and may even intensify.

12.1.2 Agriculture

Changes in agricultural practices in Scotland have been continuous since the 19th century. Many of these changes are associated with land improvement and the consequent increase in arable land. These changes involve major land drainage schemes, which generally result in lower dry weather flows and increased rates of run-off in periods of heavy rainfall. In addition, any increase in the amount of arable land leaves more bare earth exposed during the autumn and winter. As a consequence, larger quantities of top soil are carried into water courses in the run-off following heavy rain or snow melt and as a result of wind-blow when the soil is sufficiently dry. This fine silt quickly drops out of suspension, burying the gravel and with it the juvenile rearing habitat. However, since 1985, there has been a dramatic downturn in the implementation of new drainage schemes. This has come about as a result of a general lack of confidence in the farming industry coupled with a reduction in grant rate to 15% for land classified as low ground and to 30% for less favoured land. As a result, future drainage effort will be concentrated on the most productive land to maintain high levels of production there. Furthermore, as farmers take more and more arable land out of cereal production, marked declines both in the amount of fertilizers spread on the land and in the amount of chemical spraying necessary will result. This will lead to an improvement both in water quality and in the quality of the juvenile parr rearing habitat.

Sheep were introduced on a large scale to the Scottish uplands in the 19th century and, until the 1990s, their numbers have generally increased since then. In 1970−86, for example, sheep numbers in the Spey valley rose from 135 000 to 163 000 (Fisher & Revell 1988). Sheep cause severe damage by creating hollows on exposed river banks where they can shelter from the sun. The banks, now weakened, are washed away by the next major spate; the Water of Mark, a tributary of the R. North Esk, was on average 4 m wide in 1960 but 6 m wide by 1987 and base flows in the 1980s have declined.

Two other relatively modern farming practices which have a deleterious effect on water quality are the intensive rearing of animals for meat and milk production and the use of silage as a replacement for hay as a winter feed. The waste products, slurry and silage liquor, from these practices frequently enter watercourses during their disposal or filter into streams due to inadequate tank storage facilities, and cause serious pollution resulting in the deaths of large numbers of fish. In addition, the growing of grass for silage requires relatively heavy applications of high nitrogen fertilizers at regular intervals during the growing season with much of this fertilizer ending up in the nearest watercourse.

Other forms of disruption to catchments include more recent road building and accelerated soil erosion in association with the expanding skiing and tourist industry and an uncontrolled increase in the red deer population.

Most of these changes cause degeneration of nursery areas, and may prevent access of spawners into the burns at a critical time.

12.2 Water abstraction

Mention must also be made of water abstraction for irrigation and for industrial and domestic use. The production of hydro-power may involve the diversion of headwater streams of one river into a neighbouring river system. The construction of reservoirs, whether for the production of hydro-electric power or for potable water, always causes flooding. This can destroy substantial areas suitable for spawning and the on-growing of juvenile salmon, as does the diversion of streams, and it may increase both the quality and quantity of the habitat preferred by known predators of salmon including pike, trout and sawbill ducks. However, reservoirs, particularly those associated with public water supplies, if they are designed and operated correctly, can be beneficial because they offer the opportunity to provide adequate compensation flows throughout the year and reduce the impact of floods downstream of the reservoir. In some river systems, dry summer flows can be 50% higher than they would have been if no reservoir existed and if the statutory flow requirements had not been built into the Water Order permitting the abstraction.

The demand for potable water in Scotland has not diminished and a number of schemes requiring the abstraction of water from rivers are under active discussion at the moment. Although it is unlikely that there will be any major hydro-electric schemes in the near future, a number of minor schemes aimed at the supply of electricity to local communities is presently under consideration. Furthermore, the Salmon Act 1986 makes it possible for an individual to establish or extend a private generating station. Because the capacity of most of these stations is likely to be limited, they will probably be sited on the smaller streams where they could have an appreciable impact on the salmon populations.

12.3 Acidification

The acidification of lochs and rivers is mainly due to the emissions of the gaseous oxides of sulphur and nitrogen. When the pH drops below 5.0, a different buffering system in the soil becomes activated and deposits of aluminium become more soluble and are leached into the water. So, in addition to the direct effects of pH on fish physiology, there is the effect of aluminium, which is especially toxic to fish and other organisms at levels of pH which on their own would not be harmful. Forestry plantations tend to concentrate the acid as the sulphate particles settle on the needles and accumulate until washed off by heavy rain.

Of particular concern has been the effect of large conifer plantations, in upland regions, on the removal of sulphur dioxide from the atmosphere. This effect results from a change in air flow which the plantations bring about, plus their more direct effect by acting as collectors of air-borne particulate material which washes off into the surrounding watercourses. Cloud formation over high ground also entraps upflowing air, with its sulphur-laden water droplets, also contributing to acid rain.

Electro-fishing surveys have shown that acidic streams with a geometric mean pH of under 5.5 generally had abnormally low densities of salmonid fish and that female fish were very few or absent. Salmon were found to be more sensitive than trout (Chave 1990).

Surveys have shown that in nine rivers in England and Wales, 1050−2100 salmon (48 t) were lost in 1988 as a result of acidification. In Scotland, juvenile salmon may have been affected by acid conditions. Low densities ($<$25 fish 100 m^{-2}) have been estimated for several acid waters. The numbers lost as a result of the acidification of these waters have not been estimated and there is no evidence that smolt production has been significantly affected (Anon 1989e). In comparison, Norway reported an estimated loss of between 611 700−1 223 400 smolts in 1988 (Anon 1989e).

Most mitigation measures involve liming in some form − either adding limestone directly to the water or spreading it on the surrounding land. Swedish authorities consider that if liming had not been carried out, the potential loss of wild salmon would be about 40% of the annual total Swedish catch of wild salmon in the NASCO Convention area. So far, Sweden has spent a total of 195 million SKr (£18 million) on liming operations in its west-coast river systems. Mention was made earlier of the establishment of a gene bank in Norway to counter the threats to wild salmon stock, among which is included the effects of water acidification.

12.4 Global warming

Salmon are known to be sensitive to water temperature changes and while in the sea, 1SW fish seem to prefer warmer water than the older fish. Furthermore, the length of stay in the sea may be influenced by temperature. Therefore, it is not unreasonable to assume that any change in water temperature resulting from the 'Greenhouse effect' may have a marked influence on the habits of salmon both in fresh water and in the sea.

12.5 Summary

Provided that egg deposition is not limiting, the number of smolts which any river can produce at any one time is limited by the river's rearing capacity rather than the availability of spawning fish. The quality of the rearing habitat is not static. Changes in quality most commonly follow increased drainage which is a major requirement of many of the changes, including afforestation and modern farming practices which have occurred within watersheds. Increased drainage alters flow patterns which not only affect the immigration of spawners and the feeding of juveniles but also the quality of the habitat by shallowing watercourses and smothering spawning and nursery areas in a blanket of fine silt.

Intensive arable farming and the use of fertilizers, particularly nitrogen, herbicides and pesticides can have serious effects on river systems. These chemicals eventually leach into the watercourses and are detrimental to fish production either directly by killing the fish or indirectly through the food chain.

The abstraction of water, whether for domestic or industrial use or for irrigation can destroy juvenile habitat in the tributaries most affected and greatly reduce the rearing capacity of others where the effect has been less severe. Furthermore, the construction of reservoirs can drown substantial areas suitable for spawning and the on-growing of juvenile salmon.

Changes in the environment have had a major effect on the amount and quality of the nursery areas, with much variation between rivers. Not all stock components are likely to have been equally affected because many of the land use changes described have been confined to the upper catchment of rivers. The effects at the present time are probably the most pronounced this century.

So far, acidification has had little effect on smolt production in the British Isles compared with other countries, such as Norway.

Any change in water temperature resulting from the 'Greenhouse effect' may have a marked influence on the habits of salmon both in fresh water and in the sea.

Fluctuations in stock size can be due to changes in the physical and hydrobiological character of the river resulting from the changes in land use described. If managers wish to increase production, the first priority must be to increase the quantity of nursery areas or their quality or both. Because the marine mortality of salmon is density-independent within limits yet to be defined, the more smolts produced, the more adults there will be available to support commercial and recreational fisheries. River engineering may also be necessary to improve and maintain the holding pools so that there are always sufficient niches for the increased number of fish returning and for anglers to fish.

CHAPTER 13

THE FUTURE

13.1 The high seas fisheries

Catches in the high seas fisheries at Greenland and in the Faroes area peaked in 1971 and 1981 since when these two fisheries have operated under regulatory measures, including the operation of a TAC.

Catches at Greenland in 1986 and 1987 exceeded their TACs by 51 and 31 t but in 1988 and 1989 the fishermen failed to catch the quotas by 20 t and 563 t. The catch in 1989 (337 t) was similar to the 1983 and 1984 levels (310 and 297 t) when salmon off Greenland were scarce. However, the below average catch in 1989 was mainly due to reduced fishing effort after the initial two weeks of the season. The poor initial catches coupled with the depressed price being offered for salmon, gave fishermen every encouragement to change to exploiting other species. In addition, the Greenland government was unlikely to encourage them to continue fishing for salmon because it was costing them 20 Dkr in subsidy for every kg landed.

The Faroese catch in 1988 (the last year for which detailed data are available) totalled 243 t. This is one-third of the quota and less than half the catch taken in any year since 1979. The reasons for the low landings were that few vessels fished in November and December and that the weather in January was bad. The low catch rates from February onwards, coupled with the poor price being offered for salmon, made the fishery unprofitable. The result was that most vessels stopped fishing. Figures for catch per unit effort (the only measure of abundance of salmon) in the Faroese Economic Exclusive Zone (EEZ) do not suggest that stock abundance was significantly lower than in previous years (48.0 fish per 1000 hooks, compared with 47, 51, 36, 58 and 64 in each fishing season between 1982 and 1988).

From this scenario, it looks as though the level of catches in Greenland and Faroes will be controlled in future by economic forces rather than by quota (assuming that the production of farmed salmon does not fall below its present level). Therefore, it would be wasteful to spend the sum of money (£2 million) which it is presently thought can be raised to buy out the Faroese and Greenland quotas. This money should be spent on research directed at providing the data to answer the same questions which NASCO pose each year and for which there are presently no answers because the necessary information is lacking. However, there is just the possibility that the Faroese scene may change due to a significant increase in the number of catchable salmon in the Faroese fishery area if the proposed Norwegian ranching project, whereby it is intended initially to release 2 million smolts annually, reaches fruition.

13.2 Home water catches

At home, the salmon netting industry has declined and the rate of decline has accelerated in the last ten years. As a result, the total Scottish salmon catch in 1989 was the smallest since official records began in 1952 and the 1990 catch is likely to be even less. Although this decline was associated with a decrease in net fishing effort and the inability of anglers to harvest the increased size of the proportion of the stock now available to them, these are not the only reasons for the decline. An analysis of historical catch records shows that widely fluctuating catches have been characteristic of the Scottish scene, at least since the early 1800s, and that depressed catches were associated with changes in the timing of salmon runs. These changes were towards a later arrival in home waters of most of the stock, mainly because of an increase in the proportion of grilse and a decrease in the proportion of the older sea age groups which returned before and during the first half of the fishing season.

Because grilse do not appear in catches in any number until late June or early July, and because the closing dates for most net fisheries are towards the end of August, any factor delaying the return of these fish and their migration into fresh water can drastically reduce the proportion of the total stock available to the net fishery. Delay in return is unlikely to occur when MSW fish are plentiful and form the backbone of the catch. This is because MSW fish tend to migrate over a much longer period, beginning in the later months of the previous year and continuing after the end of the fishing season. Data from the automatic fish counter on the R. North Esk have shown that in some years more than 60% of the spawning stock can migrate upstream of Logie after the end of the fishing season. This is one of the main reasons why catch is an unreliable index of stock. Delay in return has less effect on rod catches because angling continues for up to three months after net fishing has ended.

Many district salmon fishery boards should take a more enlightened approach to management than at present. In any district where salmon runs have become later, application should be made to the Secretary of State to make an annual close time Order, fixing new dates between which fishing is permissible.

13.3 The demise of the net fishery

During the last 150 years, the number of active netting stations in Scottish rivers and along the coast has changed frequently. In the first half of the 19th century, for example, the amount of commercial fishing in rivers declined and was replaced by angling. Because the number of anglers, particularly from England, continued to increase, the owners of rod fisheries were pressurized into reducing still further the number of active net fisheries, especially the fixed engines associated with river estuaries. The present buy-out of nets is a repeat of similar action taken approximately 100 years previously and seems to have the similar objectives, of increasing the rod and line catch and the capital value of rod fisheries. The conservation of salmon does not enter the equation.

An added incentive to buy out nets at the present time is the depressed state of the net fishing industry. This results from the impact of farmed fish on the market price of wild salmon and from the severe statutory restraints under which the fishery is forced to operate. This situation produces willing sellers. In addition, there is a good demand and price for rod fishings, particularly those leased or sold with a no-netting tag.

Before 1980, the price which net fishermen obtained for their salmon reflected the availability of the fish. Net fishermen could usually weather periods of low catches until the timing of the runs changed. This is no longer the case because the price of salmon is now almost entirely governed by farmed salmon production. The tonnages which the fish farms decide to expose for sale can affect the market at any one time. As a result, the price of salmon, including that for wild salmon, in summer when most of the catch is taken has decreased dramatically although there was some recovery in 1990.

The salmon fishing industry is labour intensive. At present, labour charges account for more than 50% of the total operating costs. If the industry is to survive, these costs will have to be reduced. Some pruning of costs has already occurred, in particular the closure of the less profitable netting stations and delaying the opening of the season at most of the surviving net fisheries. However, this action has been insufficient to arrest the economic decline. The one remaining course open to the netsmen is an urgent evaluation of their present gear. The gear presently in use and the mode of its operation have changed little over the years because the fishermen's target has always been to produce good quality fish even at the cost of inefficiency.

To put the scenario into context, a netsman has presently to catch about three times as many salmon as previously to have the same profit margin. Because the level of exploitation by nets is relatively low, it is not possible to increase the catch sufficiently to offset the real decrease in the market price of salmon.

In the past, the traditional Scottish salmon netsmen shunned the use of any enmeshing gear because it damaged the catch and an unacceptable number of fish dropped out before it could be removed by the fishermen. Modern twines have largely reduced both these unacceptable factors so perhaps now is the time to reconsider their stance. If fished within headlands, the majority of the catch would have originated in the adjacent rivers and be little different from that taken by net and coble in the corresponding estuaries. An annual TAC for each sea age group could be fixed before the start of each season and the number of fish migrating upstream could be monitored using an automatic fish counter. A change in the law would be necessary, but in the field of salmon conservation which has been associated with the presence of an active net fishery for at least 150 years, the advantages of such a change would far outweigh any short-term disadvantage. Fixed gill nets are presently used in a number of countries including Canada and no problems have been reported.

Each net would be a fraction of the cost of a present fixed engine. Their use would introduce a measure of flexibility into the industry, require less labour to operate and be less prone to storm damage. Less capital would be tied up in the net store, where presently it is necessary to have three nets for each one fishing. There should be no necessity to fish more nets or to catch more fish than previously to show a profit because costs would be considerably reduced by the change to fixed gill nets. In addition, the valuable role played by each netsman or guardian of the salmon on the coast would continue. Furthermore, because the gear is so easily set and removed, the conservation of salmon stocks could realistically be based on a policy of controlled exploitation in the relevant fisheries to ensure an adequate spawning population.

Government must also act positively to help preserve an industry which has played such a vital role in the management of salmon stocks, through the membership of district salmon fishery boards, and other conservation bodies. Government could relax some of the draconian statutory measures, e.g. the increase in the weekly close time for nets by 18 h recently enacted and for which there was no justification on conservation grounds, as instanced by the results of research published by Freshwater Fisheries Laboratory, Pitlochry staff. In this context, the recent attitude which government has adopted, after a most exhaustive review, by their own scientists, of salmon net fishing in the areas of the Yorkshire and Northumbrian regions of the NRA and the salmon fishery districts from the R. Tweed to the R. Ugie, is encouraging. They intend to invite the NRA to consider how the fisheries in Yorkshire and Northumbria can be changed to increase the opportunities for inshore fishing by 'T' and 'J' nets (Anon 1991). Both these gears are fixed engines, which are considerably cheaper to construct and maintain than Scottish bag and stake nets (fixed engines) but as they rely for much of their catch on their ability to enmesh fish, their use in Scotland, in common with fixed gill nets, is presently illegal.

It is, perhaps, best to return to the common sense of W.J.M. Menzies (1949):

'It is equally certain that the primary object of all those concerned with fishery administration should be the production of the greatest possible stock of fish and, the abstraction from that stock, of the greatest possible number which can be spared without endangering the stock of future years.'

13.4 The possible effects of the removal of net fisheries

It cannot be assumed in any fishery district, that the total net catch will automatically become available to the rod fishery if the nets stop fishing. The most realistic assessment suggests that just over half will become available and then only during the second half of the fishing season and mostly to the rods fishing the lowermost beats because the majority of the extra fish will be grilse (Appendix C).

Some measure of the effect of the closure of net fisheries on the rod and line catch can be obtained more directly from an analysis of angling catches taken before and after net fishing ceased. On the Aberdeenshire R. Dee, for example, the netting station operated by the River Dee Improvement Association ceased fishing in 1968 leaving the one fishery owned by the Aberdeen Harbour Board operating downstream until it also closed in 1986. Mean annual rod catches of spring salmon, total salmon and grilse for the ten years following the closure were 3174, 5205 and 187 compared with 5878, 9688 and 179 for the 10 years prior to the closure. Salmon rod catches in 1987 and 1988 following the closure of the remaining net fishery have not changed significantly but grilse catches are the highest recorded since records began in 1952. Part of this increase was due to increased angling effort on the beats nearest to the sea in the autumn. Previously, many beats stopped fishing in June.

There may be no increase in rod catch until banks are repaired and the gravelling of pools, which has dramatically increased in recent years, is arrested and the holding capacity of these pools is restored to their previous levels and maintained.

If optimum egg deposition can be obtained despite a net fishery, increase in egg deposition following the removal of that fishery may not result in more smolts until either the quality or quantity of nursery areas, or both, are increased. However, the stock of fish in a river may consist of a number of components, e.g. spring and summer salmon and grilse. It is possible that while the overall egg deposition is at or above the optimum level, the egg deposition of one or more of the stock components could be below this level. In this case, the removal of the net fishery may protect the scarcer components by allowing sufficient additional fish to spawn to raise their level of egg deposition to or above the optimum level. Conversely, if the cessation of netting merely allows an increase in the number of spawners in any plentiful component of the stock, these fish could depress still further the weaker components in the stock by, for example, overcutting redds already constructed. The scarcest stock component at present is the spring fish. Because fish belonging to this group tend to spawn earliest in each autumn, overcutting by later spawning fish may be a real danger.

Nowadays, net fisheries usually fish only in summer when grilse enter into the fishery. The prohibition of netting is likely to increase the number of grilse spawners. This is the stock component which appears to be sufficiently strong at present to seed the available spawning areas adequately.

In Scotland, most spring fish are female and most grilse are male. Any addition in the number of grilse could increase the possibility of spring fish and grilse mating. The resultant cross might not possess the gene complex required to increase the proportional return in the spring. Thus, the number of potential spring fish would be further reduced.

If the number of fish in the river is increased, particularly in the summer when they are unable to disperse because of low flows, ideal conditions are created for the spread of any disease present. Mortality could then increase to such a level that the number of fish actually spawning could be less than the previous minimum in the presence of an active net fishery. In addition, there is some evidence to suggest that after the cessation of a legal net fishery, the removal of fish by illegal methods increases and may exceed what would have been taken by the legal fishery. The increased number of illegal nets reported to have been confiscated in areas where salmon fisheries once operated by the Fishery Protection Agency tend to support this suggestion. In addition, seals can now feed unhindered along large tracts of coast and establish new breeding colonies. Such an event occurred at Catterline, north of Montrose, relatively soon after the netting station closed and following a marked increase in the number of seals observed in the vicinity.

The demise of the net fishery has another serious implication in that an extremely useful management tool may be lost.

The total exploitation by fisheries on the smolts which emigrated to sea from the R. North Esk in 1980−85 was 73−88%. These levels of exploitation were sustainable. As angling alone cannot achieve these levels of exploitation, the consequence of removing the nets from the R. North Esk would be a gross excess of spawners with many associated problems.

Most salmon biologists agree that a river can only produce a finite number of smolts each year. Where egg deposition is not a limiting factor, this number is controlled by density-dependent mortality on the young stages of the fish. Once the optimum number of eggs has been deposited, additional eggs will not result in more smolts. However, the optimum number of eggs can be increased by increasing the quantity of nursery areas, or their quality, or both (Fig. 13.1). The optimum number can be reduced by destroying nursery areas. The river may, therefore, always produce the maximum number of smolts for the current conditions.

Thus, there is no substance whatsoever in the idea that although the absence of an active net fishery is of marginal benefit to the rods, the stock will benefit enormously because smolt production will be enhanced by several orders of magnitude.

On the R Tyne, rod catches have increased steadily since the mid-1960s in the presence of an active off-shore drift net fishery and a fixed net fishery along the coast. These data show that net fisheries do not necessarily depress the number of fish caught by rod and line. Thus, it should not be assumed that rod catches will increase automatically following the closure of net fishing.

13.5 Seals

Predation by both grey and common seals will remain a problem. Common seals in particular have been regularly observed feeding miles up river in the Rivers Tweed, South Esk, Dee, Spey, Beauly and Naver, and in the estuary of the R. North Esk. As their numbers continue to increase, so will their share of the resource. This seems likely to continue until such time as an increase in

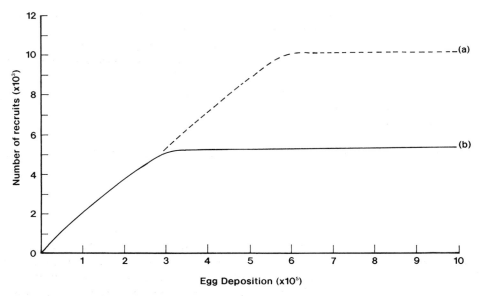

Fig. 13.1 Salmon stock−recruitment relationship before (a) and after (b) the improvement and/or the increase in nursery area.

natural mortality intervenes, causing their numbers to drop dramatically. It is unlikely that permission to cull the population will be granted again in the foreseeable future. Even if such a will existed, the logistics of a meaningful cull, in terms of number, are such as to rule it out immediately, solely on the grounds of practicability. Seals can be shot in Scotland outwith the breeding season, and licences may be granted to shoot grey and common seals during their breeding seasons. The Norwegians have adopted a more positive attitude and have actively culled their populations of seals to conserve fish stocks in the interests of their fishermen. This action is probably easier to defend in Norway because fish are much more important to their economy than they are in Scotland.

13.6 Research

Although major advances in our knowledge of salmon biology have occurred in the last 30 years, major gaps still remain. The areas where future research should concentrate include the following:

(1) Juvenile habitat preferences,
(2) The interaction between inherited characteristics and environmental factors in relation to the number of years spent in the river before smolting, the number of winters subsequently spent in the sea, the month of return and the preferred spawning site of each age group,

(3) The relationship between rod catch, angling effort (including fishing efficiency), and the number of fish in the river,
(4) The effect of the removal of net fisheries on smolt production,
(5) The impact of aquaculture on wild salmon stocks, with particular reference to escapees from fish farms spawning in local rivers,
(6) The movements of salmon in the North Atlantic particularly in relation to their feeding grounds,
(7) Marine survival.

It should be noted that the above list is neither exhaustive nor in order of importance.

13.6.1 *Juvenile habitat preferences*

Garcia de Leaniz (1990) suggested that young salmon in the Girnock Burn are not territorial and do not distribute themselves evenly over the available habitat. They are found clumped at all juvenile densities around particular sites and much of the burn is barren. Juvenile salmon, therefore, select some sites and avoid others. Further research is required to describe the sites selected in great detail and whether it would be possible to increase their number by, for example, river engineering. If successful, it should be possible to increase the rearing capacity per unit area and, as a result, the number of smolts produced without altering their age structure. A 1% reduction in mortality at the alevin stage doubles the number of potential smolts.

13.6.2 *The interaction between inherited characters and environmental factors*

At present, catches of MSW fish have declined, particularly the fraction caught in the first half of the fishing season. Catches of grilse have increased and as a result there has been a marked increase in the proportion of the total stock returning after the end of the fishing season. The results of research carried out in the R. North Esk showed that:

(1) There was a relationship between smolt age and time of return within each sea age group,
(2) The age structure of the juveniles differed between burns but not with the adults spawning in the burn,
(3) The increased percentage of grilse in the catch coincided with below average temperatures in the sub-arctic,
(4) Fish returning to fresh water at different times of year spawned in different areas of the watershed. It is not known whether each of these distinct groups of fish are genetic strains tending to breed other similar groups or whether the inherited characteristics can be influenced by environmental factors.

It would be a major step forward if it was known whether a spring run could be recreated in an upland tributary by seeding that tributary with eggs derived from late running fish which are presently the dominant stock component.

13.6.3 *The relationship between rod catch and the number of fish in the river*

Unfortunately, our knowledge of the relationship between the numbers of fish and rod catch is extremely limited because of a reluctance of anglers in the past to co-operate in research, provide material and release their catch figures. What evidence there is suggests that the relationship between catch and the number of fish is not a simple linear relationship (Mills *et al.* 1986). It suggests that there is an inverse relationship between numbers and catch. Relatively recent data from Canada, Iceland, England and the R. North Esk have shown that when stock levels are low, the exploitation rate by rod and line is above average and at high stock levels, the overall catchability of fish declines. The limited data available following the recent closure of net fisheries tend to support this view, since rod catches have not always increased following the removal of the nets. Furthermore, the results of tagging experiments in the R. Spey indicate that catchability changes with time since leaving the sea (Shearer 1988c). It is not clear, however, whether all sea age groups have the same catchability coefficient and whether increasing angling effort will result in higher catches.

13.6.4 *The effect of the removal of net fisheries on smolt production*

Because the buy-out of nets may continue, the need for research in this field is urgent if only to monitor the effect of the increase in the number of spawners on smolt production and the age structure of the emigrating smolts.

13.6.5 *The impact of aquaculture on wild salmon stocks*

The development of salmon farming has led to an increased number of reared fish in the wild. Their impact on wild stocks can take a number of forms through ecological and reproductive interactions. If, for example, these fish had escaped from a marine site, they would not be imprinted to a home stream and would stray into many rivers. In these rivers they will mate with other farmed or wild fish.

The impact of interbreeding between wild and farmed salmon will be to introduce into the wild population genetic types which would normally be infrequent or absent. As the genetic make-up of wild populations will have been moulded by natural selection to provide the mix of genetic types which is optimal for long term survival and maximum productivity, the new genetic types would, on average, be expected to be less well adapted. If so, then the productivity of the population will decrease.

In the autumn of 1989, fish farm escapees were observed spawning in a Scottish river in quantity, so resources require to be allocated to investigate and monitor the effect which similar occurrences may have on productivity. The key areas are smolt production, sea survival, the number of adults returning from each spawning and when and to where they return.

13.6.6 *The movements of salmon in the North Atlantic*

Prior to about 1959 and the birth of the high seas fisheries, the exploitation of salmon was largely in the hands of the country of origin and as a result, each country was free to legislate accordingly. There was also little incentive and even less encouragement to investigate the marine phase of the life history. However, this situation changed almost overnight with the development of the high seas fisheries. Apart from the immediate international management requirements, the presence of these fisheries stimulated research into assessing the long and short term effects of high seas fisheries on stocks already exploited in home waters.

Run reconstruction modelling has recently been used by the North Atlantic Salmon Working Group, set up under the auspices of ICES principally to advise NASCO, to describe the ways in which the numerical strength of a cohort of smolts declines between immigration from the river and return as adult salmon after spending one or more winters in the sea. The model uses tagging data from index rivers like the R. North Esk where there is also information on spawning escapement, the number of immigrating smolts and the numbers of fish caught in the various fisheries either as real catches or estimates. Additional input data required include the timing of the fisheries exploiting R. North Esk stock and the proportion of the non-maturing stock which migrates to west Greenland. Given these data or estimates, the model derives the exploitation level for each of the fisheries on each adult age-group and the proportion of fish in each adult age-group which mature and return to homewaters.

The model, together with an example using data from the R. North Esk, is described by Shelton & Dunkley (in press).

The deficiencies curtailing the development of this and other models to assess the contribution which individual stocks make to each fishery and the exploitation rates on stocks in these fisheries include the lack of knowledge of migratory routes, the movement of fish between the different high seas fisheries, and the fraction of a stock available to high seas fisheries. Tagging at sea is difficult and costly but such an investigation would seem to be essential to further the development of models to a stage where they would be of practical use to manage a fishery.

13.6.7 *Marine mortality*

Marine survival has been discussed by the North Atlantic Salmon Working Group on many occasions. In 1984, ICES requested this Group to assess possible causes of the apparently poor marine survival of salmon stocks contributing to many, but not all, fisheries on both sides of the Atlantic in 1983.

The result of their deliberations was that there was no information to identify possible causes, but observations in Sweden and Iceland suggested that marine survival of salmon can be adversely affected by cold temperatures at sea (Anon 1984a).

Almost ten years later, the causes of low catches in distant and home water fisheries are again being attributed to high mortality at sea, but identification of the real cause is not any closer. Marine research involving the commissioning of a vessel capable of remaining at sea for several weeks at a time and having the capability of fishing various types of gear on the high seas is costly.

In 1985, the Working Group estimated that the annual cost of catching and tagging salmon off the Faroes would be between £306 000 and £1 450 000. The lower cost was based on an optimistically high tag-return rate. This is the sort of money which will have to be spent annually if we intend to address the marine mortality problem. However, this sum should not be considered in isolation but in the context of the value of salmon to the economy of each country and the capital value of a single fish, which, to the owner of a rod fishery, in Scotland is reckoned to be £17 000.

13.7 Management requirements

Over 38 million juvenile Atlantic salmon at various stages may be stocked annually into rivers draining into the North Atlantic (Kennedy 1988). However, there is a great disparity of view between various managers, not just concerning the efficiency of different stocking methods, but also the value of enhancement *per se*. These differences arise mainly because of the paucity of data available to quantify the returns from enhancement projects. In most instances, the experiment is assumed to have been completed once the juveniles have been released. This is not a management practice which should be followed. A comprehensive monitoring programme should commence following each enhancement project.

Since the inception of the high seas fisheries, it has no longer been possible to manage salmon stocks on a national basis. However, much of the basic data necessary are the same whether the stocks are managed at the international, national or river-by-river level, because the objectives are broadly similar. These are the conservation of stocks and the optimization of yield. A third objective, because of its socio-economic implications, is the minimization of the variability in catches from year to year. The success of any management plan would be the ability to attain the predetermined target of each sea age or stock component each year while still fishery.

In setting a target number of spawners having the same year there will be wide annual fluctuations in the producti due to varying levels of natural mortality in fresh water annual production of adult recruits to the fisheries probabl $\pm 30\%$ from the average for a given egg deposition (Anon 1 strategies are available for surmounting this problem, but the one presently favoured is that of fixing the harvest in mixed stock fisheries, e.g. the high seas fisheries, at a sufficiently low level to allow final adjustments to the spawning escapement of each stock component in or close to the river of origin by regulating fishing pressure. Unfortunately, in Scotland, this most valuable management tool may not be available for much longer because of the closure of net fisheries.

Many district salmon fishery boards elected to manage salmon rivers and their migratory fish stocks have decided that there is a real need to construct and implement management plans pertinent to their rivers; accordingly, a small number (4) have employed biologists. For those which do not wish to employ biologists, there is another option, the employment of a part-time consultant. Such consultants have all the expertise and facilities to collect and analyse the data required to construct a management plan, in consultation with the relevant district salmon fishery board, advise on its implementation, and monitor whether the objectives are being attained.

The values or reliable estimates which are required to construct a management plan and monitor the results are described in Appendix D. Once these data are available, catches and fishing seasons can be manipulated, for example, by varying fishing effort by the different gears and the duration of the weekly and annual close times to maximize the returning stocks for the benefit of both the net and rod fisheries, at the same time as attaining a sufficient escapement of each component in the stock to seed the respective juvenile nursery areas at their maximum rearing capacity. Furthermore, nursery areas could be managed in such a manner that unit production would be maintained without necessarily altering the age structure of the juveniles produced. Stream sections inaccessible to adults could be brought into production by the removal of existing barriers.

Because it is unlikely that density-dependent mortality operates at sea, the number of fish returning will be correlated with the number of smolts produced. Use of a data base allows all management strategies to be monitored and assessed.

Natural reproduction at low stock densities is a very efficient process, and the management of the number of fish which return to spawn is probably the most effective means of enhancing depleted stocks. At higher stock levels, improvements to nursery areas to increase the number of fish reared to the smolt stage might be most effective, e.g. river engineering to improve access or regulation of discharge to avoid becoming devoid of water (Solomon 1984).

scheme developed to enhance the stock must be monitored and the
est method of obtaining the necessary data is to count the number of
ending adults because one has little or no control over what happens
outwith the river. To do this, it is absolutely essential to instal an automatic
fish counter in the river together with a fish trap which will allow both the
incoming and outgoing stock to be sampled.

13.8 Summary

The level of catches in the Greenland and Faroes fisheries will be controlled
in future by economic forces rather than TACs.

Scottish salmon catches have declined for reasons other than the decline in
the size of the overall river stock. These include, the decrease in netting effort,
the inability of angling to increase the harvest to replace the deficit left by
the decline in the net fishery and the increased proportion of the stock which
is no longer catchable because it returns to home waters outwith the fishing
season. Historical records suggest that the timing of the runs can change and
thereby decrease the proportion of the total stock available for exploitation by
the different gears.

In the historical records, there is little evidence of a major run of spring
salmon in the 19th century in the larger Scottish east coast rivers.

The most realistic assessment of the immediate effects of the removal of a
net fishery indicates that additional fish will be available to the rods but the
rod catch will only increase provided the catchability of the fish does not
decrease. Even if that occurs, the beats furthest from the sea would reap little
benefit.

It is unlikely that smolt production would increase. In fact there is a real
danger that it might be depressed, particularly the smolts derived from the
eggs deposited by the early running fish.

There is little doubt that seals (mostly grey) are consuming an increasing
number of salmon which would otherwise be available to fisheries and for
spawning. It is unlikely that any government, irrespective of its politics, will
have the will to sanction a cull of sufficient magnitude to have a measurable
effect on the level of catches and spawning stocks, even if the logistic problems
associated with such action could be solved. Therefore, fishermen should
make full use of the concessions granted to them to protect their fisheries.

Because of the rapidly changing scene both at home, including the salmon
farming industry, and in the high seas fisheries, there is an urgent need to
commission more research. A number of projects are described and the significance of each is examined.

A monitoring programme is discussed in the context of the need to manage
stocks on a scientific basis.

BIBLIOGRAPHY

Adams D.C. (1985) Report on the history of salmon fishing in the Montrose area, c 1360—1835. *Montrose Library*, **58**, 639, 2755.

Alabaster J.S. (1970) River flow and upstream movement and catch of migratory salmonids. *J. Fish Biol*, **2**, 1—13.

Allan I.R.H. & Ritter J.A. (1975) Salmonid terminology. *J. Cons. Perm. Int. Explor. Mer.*, **37**, 293—9.

Allen K.R. (1941) Studies on the biology of the early stages of the salmon (*Salmo salar*). 2 Feeding habits. *J. Anim. Ecol.*, **10**, 47—76.

Altukhov Yu. P. (1981) The stock concept from the viewpoint of population genetics. *Can. J. Fish. Aquat. Sci.*, **38**, 1523—8.

Andersen K.P., Horsted Sv. Aa. & Møller Jensen J. (1980) Analysis of recaptures from the West Greenland tagging experiments. *Rapp. P.-v. Réun. Cons. int. Explor. Mer.*, **176**, 136—41.

Anderson J.M. (1986) Merganser predation and its impact on Atlantic salmon stocks in the Restigouche river system, 1982—1985. *Atlantic Salmon Federation Special Publication Series*, **13**, 66 pp.

Anon (1932) Annual Report Fishery Board for Scotland, 1931. *Her Majesty's Stationery Office, London.*

Anon (1961) Report of committee on salmon and freshwater fisheries. *Her Majesty's Stationery Office, London*, **Cmnd. 1350.** (The Bledisloe Report.)

Anon (1963) Scottish Salmon and Trout Fisheries. *Her Majesty's Stationery Office, London*, **Cmnd. 2096.** (First Hunter Committee Report.)

Anon (1964) Laerebog i fangst for Syd-og Nordgronland. Den kongelige gronlandske Handel, København 1964. (In Danish.)

Anon (1965) Scottish Salmon and Trout Fisheries. *Her Majesty's Stationery Office, London*, **Cmnd. 2691.** (Second Hunter Committee Report.)

Anon (1974) Taking stock. A report to the Association of River Authorities. *Association of River Authorities March 1974*, 48 pp.

Anon (1981) Report of the Working Group on North Atlantic Salmon, 1979 and 1980. *International Council for the Exploration of the Sea, Copenhagen*, Cooperative Research Report No. **104**, 46 pp.

Anon (1983) Salmon Conservation — A New Approach. Report of the Salmon Sales Group of the National Water Council. *National Water Council*, July 1983, 31 pp.

Anon (1984a) Report of the Working Group on North Atlantic Salmon. *International Council for the Exploration of the Sea, Copenhagen*, **C.M.1984/Assess:16**.

Anon (1984b) The impact of grey and common seals on North Sea resources. *Natural Environment Research Council Sea Mammal Research Unit, Cambridge*, Contract no. ENV 665 UK (H) — Final Report, 150 pp.

Anon (1984c) A study of the economic value of sporting salmon fishing in three areas of Scotland. *Tourism and Recreation Research Unit, University of Edinburgh*, 21 pp. (The TRRU Report.)

Anon (1985) Report of the Working Group on North Atlantic Salmon. *International Council for the Exploration of the Sea, Copenhagen*, **C.M.1985/Assess:11**.

Anon (1986a) Report of the Working Group on North Atlantic Salmon. *International Council for the Exploration of the Sea, Copenhagen*, **C.M.1986/Assess:17**.

Anon (1986b) Report of the meeting of the Special Study Group on the Norwegian Sea and Faroese Salmon Fishery. *International Council for the Exploration of the Sea, Copenhagen*, **C.M.1986/M:8**.

Anon (1986c) Investigation into the alleged decline of migratory salmonids, final report. *Wessex Water Authority, Bristol*, 40 pp.

Anon (1987a) Report of the Study Group on the Norwegian sea and Faroese Salmon Fishery. *International Council for the Exploration of the Sea, Copenhagen*, **C.M.1987/M:2**.

Anon (1987b) Report of the Salmon Review Group. Framework for the development of Ireland's salmon fishery. *Stationery Office, Dublin*, 103 pp.

Anon (1987c) Report of the Working Group on North Atlantic Salmon. *International Council for the Exploration of the Sea, Copenhagen*, **C.M.1987/Assess:12**.

Anon (1988) Report of the Working Group on North Atlantic Salmon. *International Council for the Exploration of the Sea, Copenhagen*, **C.M.1988/Assess:16**.

Anon (1989a) Annual Review 1986−1987. *Freshwater Fisheries Laboratory, Faskally, Pitlochry*, 35 pp.

Anon (1989b) Annual Review 1987−1988. *Freshwater Fisheries Laboratory, Faskally, Pitlochry*, 35 pp.

Anon (1989c) Report of the Working Group on North Atlantic Salmon. *International Council for the Exploration of the Sea. Copenhagen*, **C.M.1989/Assess:12**.

Anon (1989d) Report of the Study Group on Toxicological Mechanisms Involved in the Impact of Acid Rain and its Effects on Salmon. *International Council for the Exploration of the Sea, Copenhagen*, **C:M.1989/M:4**.

Anon (1990a) Report of the North Atlantic Salmon Working Group. *International Council for the Exploration of the Sea, Copenhagen*, **C.M.1990/Assess:11**.

Anon (1990b) The effects of fishing at low water levels. *The Salmon Advisory Committee MAFF Publications, SE997TP*, 17 pp.

Archer W.E. (1895) Report on food of salmon. *Fourteenth Ann. Rep. Fish. Bd. Scotland., Scot. Salm. Fish*, 1895, 23−9.

Baker R. (1978) *The evolutionary ecology of animal migration*. Hodder and Stoughton, London.

Balmain K.H. & Shearer W.M. (1956) Records of salmon and sea trout caught at sea. *Freshwat. Salm. Fish. Res., Scotland*, **11**, 12 pp.

Berg M.M. (1977) Tagging of migrating smolts (*Salmo salar*, L.) in the Vardnes River, Troms, Northern Norway. *Rep. Inst. Freshwat. Res., Drottningholm*, **56**, 5−11.

Bertmar G. & Tofr R. (1969) Sensory mechanisms of homing in salmonid fish I. Introductory experiments on the olfactory sense in grilse of Baltic salmon. *Behaviour*, **35**, 234−41.

Beverton R.J.H & Holt S.J. (1957) On the dynamics of exploited fish populations. *Fishery Invest., Lond. Series 2*, **19**, 1−533.

Boece H. (1527) A History of Scotland (*Scotorum Historiae: Scotorum Regni Descriptio, folio XII*), Paris.

Brown D.W. (1981) The design and construction of an experimental fish counter in the River North Esk, at Logie, near Montrose. *Civil Engineering and Water Services, Scottish Development Department, Edinburgh*, 20 pp.

Browne J. (1986) The data available for analyses on the Irish salmon stock. In *The Status of the Atlantic Salmon in Scotland* (Ed. by D. Jenkins & W.M. Shearer), 84−90 (*ITE Symposium no. 15*). Abbots Ripton: Institute of Terrestrial Ecology.

Browne J., Eriksson C., Hansen L.P., Larsson P.-O., Lecomte J., Piggins D.J., Prouzet P., Ramos A., Sumari O., Thorpe J.E. & Toivonen J. (1983) *C.O.S.T. 46, Mariculture, XII/ 1199/82, E.E.C., Brussels*, 94 pp.

Buck R.J.G. & Hay D.W. (1984) The relation between stock size and progeny of Atlantic salmon (*Salmo salar* L.) in a Scottish stream. *J. Fish Biol*, **24**, 1−11.

Buckland F. (1880) *Natural History of British Fishes*, SPCK, Unwin Bros, London.

Calder A.M. & Gill J.G.S. (1988) Forestry on Speyside: its evolution and production. In *Land Use in the River Spey Catchment*, (Ed. by D. Jenkins), 128−133. (ACLU Symposium no. 1). Aberdeen: Aberdeen Centre for Land Use.

Calderwood W.L. (1907) *The Life of the Salmon*. Edward Arnold, London.

Calderwood W.L. (1930) *Salmon and Sea Trout*. Edward Arnold, London.

Calderwood W.L. (1940) Thirty years of salmon marking. *Salmon and Trout Magazine*, **88**, 207−13.

Camino E.G. (1929) The sea life of Spanish salmon. *Fish Gaz*, **99**, 422.

Chadwick E.M.P. (1985) The influence of spawning stock on production and yield of Atlantic salmon (*Salmo salar* L.) in Canadian rivers. *Aquaculture and Fisheries Management*, **16 (1)**, 111−9.

Chave P. (1990) Waterways of the future: the role of biological assessment in protecting our aquatic environment. *Biologist*, **37 (4)**, 129−31.

Child A.R. (1980) Identification of stocks of Atlantic salmon (*Salmo salar* L.) by electrophoretic analysis of serum proteins. *Rapp. P.-v. Réun. Cons. int. Explor. Mer*, **176**, 65−7.

Clark H.J.S. (1952) Salmon fishing and the weir at Wareham. *Salmon and Trout Magazine*, September 1952.

Coull J.R. (1979) Fisheries in the rivers and lochs of Scotland in the 16th, 17th and 18th centuries: The evidence from MacFarlane's Geographical Collections. *Salmon Net*, XII, 46−54.

Cushing D. (1983) *Climate and Fisheries*. Academic Press, London.

Doubleday W.G., Rivard D.R., Ritter J.A. & Vickers K.U. (1979) Natural mortality rate estimates for North Atlantic salmon in the sea. *International Council for the Exploration of the Sea, Copenhagen*, **C.M.1979/M:26**.

Døving K.B., Johnsson B. & Hansen L.P. (1984) The effect of anosmia on the migration of Atlantic salmon smolts (*Salmo salar* L.) in freshwater. *Aquaculture*, **38**, 383−86.

Dunbar M.J. (1981) Twentieth century marine climatic change in the northwest Atlantic and subarctic regions. In *Symposium on Environment Conditions in the northwest Atlantic during 1970−79*, 7−15, NAFO Scientific Council Studies 5, Dartmouth, Nova Scotia, Canada.

Dunbar M.J. & Thomson D.H. (1979) West Greenland salmon and climatic change. *Meddelelser om Grønland*, **2022**, 1−19.

Dunfield R.W. (1985) The Atlantic salmon in the history of North America. *Canadian Special Publication of Fisheries and Aquatic Sciences*, **80**, 181 pp.

Dunkley D.A. (1986) Changes in the timing and biology of salmon runs. In *The Status of the Atlantic Salmon in Scotland* (Ed. by D. Jenkins and W.M. Shearer), pp. 20−27 (*ITE Symposium No. 15*). Abbots Ripton: Institute of Terrestrial Ecology.

Dunkley D.A. & Shearer W.M. (1982) An assessment of the performance of a resistivity fish counter. *J. Fish Biol*, **20**, 717−37.

Egglishaw H.J. (1967) The food, growth and population structure of salmon and trout in two streams in the Scottish Highlands. *Freshwat. Salm. Fish. Res*, **38**, Scotland, 32 pp.

Egglishaw H.J. (1970) Production of salmon and trout in a stream in Scotland. *J. Fish Biol*, **2**, 117−236.

Egidius E. (1987) Import of furunculosis to Norway with Atlantic salmon smolts from Scotland. *International Council for the Exploration of the Sea*, **C.M.1987/S:8**.

Elson P.F. (1975) Atlantic salmon rivers smolt production and optimal spawning: an overview of natural production. *International Atlantic Salmon Foundation, Special Publication Series*, **6**, 96—119.

Elson P.F. & Tuomi A.L.W. (1975) *The Foyle Fisheries: New Bases for Rational Management*. The Foyle Fisheries Commission, Londonderry, Northern Ireland.

Fisher J. & Revell B.J. (1988) Agricultural development and the regional economy of the Spey Valley. In *Land Use in the River Spey Catchment* (Ed. by D. Jenkins), pp. 116—27 (ACLU Symposium No. 1) Aberdeen: Aberdeen Centre for Land Use.

Fox I.A. (1989) The hydrology of the Tweed. In *Tweed Towards 2000* (Ed by D. Mills), pp. 29—34 (Tweed Foundation Symposium No. 1). Tweedmouth, The Tweed Foundation.

Fraser P.J. (1987) Atlantic salmon feed in Scottish coastal waters. *Aquaculture and Fisheries Management*, **18 (3)**, 243—7.

Frost W.E. (1950) The growth and food of young salmon (*Salmo salar*) and trout (*Salmo trutta*) in the River Forss, Caithness. *J. Anim. Ecol*, **19**, 147—58.

Garcia de Leaniz C. (1990) *Distribution, growth, movements and homing behaviour of juvenile Atlantic salmon and brown trout in the Girnock Burn, Aberdeenshire*. PhD thesis, University of Aberdeen.

Gardiner R. (1989) Tweed juvenile salmon and trout stocks. In *Tweed towards 2000* (Ed. by D. Mills), pp. 105—14 (Tweed Foundation Symposium No. 1). Tweedmouth: The Tweed Foundation.

Garnås E. & Hvidsten N.A. (1985) Density of Atlantic salmon (*Salmo salar*, L.) smolts in Örkla, a large river in central Norway. *Aquaculture and Fisheries Management*, **16**, 369—76.

Gee A.S. & Edwards R.W. (1981) Recreational exploitation of the Atlantic salmon in the River Wye. In *Allocation of Fishery Resources* (Ed. by J.H. Groves), pp. 129—37, (Proceedings of the Technical Consultation in Fishery Resources, Vichy, France), Rome, FAO.

Gee, A.S. & Milner, N.J. (1980) Analysis of 70-year catch statistics for Atlantic salmon (*Salmo salar*). *J. Appl. Ecol.*, 17, 41—57.

Gee A.S., Milner N.J. & Hemsworth R.J. (1978) The effect of density on mortality in juvenile Atlantic salmon (*Salmo salar*). *J. Anim. Ecol*, **47**, 495—505.

George A.F. (1982) Scottish salmon return — migration variations *c* 1790—1976. *MPhil Thesis, The Open University, Milton Keynes*.

Gibson R.J. & Côté Y. (1985) Production de saumoneaux et récaptures de saumons adults étiquetés à la rivière, Matamec, Côte-Nord, Golfe du Saint-Laurent, Quebec. *Naturaliste Can.*, **109**, 13—25.

Greer-Walker M., Harden Jones F.R. & Arnold G.P. (1978) The movements of plaice tracked in the open sea. *J. Cons. int. Explor. Mer*, **38**, 58—86.

Grimble A. (1902) *The salmon rivers of Scotland*. Kegan Paul, London.

Gudjónsson T. (1988) Exploitation of salmon in Iceland. In *Atlantic Salmon: Planning for the Future* (Ed. by D. Mills & D. Piggins), pp. 162—78 (The Proceedings of the Third International Atlantic Salmon Symposium, Biarritz, France). Croom Helm, London.

Hansen L.P. (1988) Status of exploitation of Atlantic salmon in Norway. In *Atlantic Salmon: Planning for the Future* (Ed. by D. Mills and D. Piggins), pp. 143—61 (The Proceedings of the Third International Atlantic Salmon Symposium, Biarritz, France). Croom Helm, London.

Hansen L.P., Jonsson B. & Andersen R. (1988) Salmon ranching experiments in the River Imsa: is homing dependent on sequential imprinting of the smolts? In *Proceedings of the Second International Symposium on Salmon and Trout Migratory Behaviour, Trondheim, June 1987* (Ed. by E. Brannon and B. Jonsson), 19—29.

Hansen L.P., Naesje T.F. & Garnås E. (1986) Stock assessment and exploitation of

Atlantic salmon *Salmo salar* L. in the River Drammenselv. *Fauna Norvegica*, Series A,**7**, 23−6.

Hansen L.P. & Pethon P. (1985) The food of the Atlantic salmon, *Salmo salar* L., caught by long-line in northern Norwegian waters. *J. Fish Biol*, **26**, 553−62.

Hansen P.M. (1965) Report on recaptures in Greenland waters of salmon tagged in rivers in America and Europe. *Int. Comm. N.W. Atlant. Fish., Redbook 1965*, Part III, 194−201.

Harden Jones F.R. (1968) *Fish Migration*. Edward Arnold, London.

Harris G.S. (1988) The status of exploitation of salmon in England and Wales. In *Atlantic Salmon: Planning for the Future* (Ed. by D. Mills and D. Piggins), pp. 169−90, (The Proceedings of the Third International Atlantic Salmon Symposium, Biarritz, France). Croom Helm, London.

Hasler A.D. (1954) Odor perception and orientation in fishes. *J. Fish. Res. Bd Can*, **11**, 107−29.

Hasler A.D. & Wisby W.J. (1951) Discrimination of stream odors by fishes and its relation to parent stream behaviour. *Amer. Nat*, **85**, 223−38.

Hawkins A.D. & Smith G.W. (1986) Radio-tracking observations on Atlantic salmon ascending the Aberdeenshire Dee. *Scottish Fisheries Research Report* **No. 36**, 24 pp.

Hawkins A.D., Urquhart G.G. & Shearer W.M. (1979a) The coastal movements of returning Atlantic salmon. *Salmo salar* L. *Scottish Fisheries Research Report* **No. 15**, 15 pp.

Hawkins A.D., Urquhart G.C. & Shearer W.M. (1979b) The coastal movements of returning Atlantic salmon. *The Salmon Net*, **XII**, 34−37.

Henderson H. (1880) *Notes and Reminiscences of my Life as an Angler*. Satchell, Leyton & Co., London.

Hislop J.R.G & Youngson A.F. (1984) A note on the stomach contents of salmon caught by long line north of the Faroe Islands in March, 1983. *International Council for the Exploration of the Sea, Copenhagen*, **C.M.1984/M:17**.

Hunt D.T. (1978) The salmon and the Crown. *The Salmon Net*, **XI**, 16−9.

Huntsman A.G. (1934) Factors influencing return of salmon from the sea. *Trans. Amer. Fish. Soc.* **64**, 351−5.

Huntsman, A.G. (1945) Freshets and Fish. *Trans. Amer. Fish. Soc.* **75**, 257−66.

Hvidsten N.A. & Møkkelgjerd P.I. (1987) Predation on salmon smolts, *Salmo salar* L., in the estuary of the River Surna, Norway. *J. Fish Biol.*, **30**, 273−80.

Ikonen E. (1987) Mixing of wild and hatchery-reared salmon during migration in the Baltic Sea. *International Council for the Exploration of the Sea, Copenhagen*, **C.M.1987/M:10**.

Jákupsstovu S.H.í. (1988) Exploitation and migration of salmon in Faroese waters. In *Atlantic Salmon: Planning for the Future* (Ed. by D. Mills and D. Piggins), pp. 458−82 (The Proceedings of the Third International Atlantic Salmon Symposium, Biarritz, France). Croom Helm, London.

Jákupsstovu S.H.í., Jorgensen P.T., Mouritsen R. & Nicolajsen A. (1985) Biological data and preliminary observations on the spatial distribution of salmon within the Faroese fishing zone in February 1985. *International Council for the Exploration of the Sea, Copenhagen*, **C.M.1985/M:30**.

Jensen A.J. & Johnsen B.O. (1986) Different adaptation strategies of Atlantic salmon (*Salmo salar*) populations to extreme climates with special reference to some cold Norwegian rivers. *Can. J. Fish. Aquat. Sci*, **43**, 980−4.

Jensen A.S. (1939) Concerning a change of climate during recent decades on the Arctic and subarctic regions, from Greenland in the west to Eurasia in the east, and contemporary biological and geophysical changes. *Biol. Meddr.*, **14**, 1−75.

Jensen A.S. (1948) Contributions to the Ichthyofauna of Greenland, 8−24, *Spolia Zool. Mus. Loun.*, **9**, 1−182.

218   *The Atlantic Salmon*

Jensen J.M. (1967) Atlantic salmon caught in the Irminger Sea. *J. Fish. Res. Bd. Can*, **24(12)**, 2639–40.

Jensen K.W. (1979) Lakseundersokelser i Eira. In *Vassdragsregulingers biologiske virkninger i magasiner og Lakseelver* (Ed. by T.B. Gunnerod & P. Mellquist), 165–71, Norge Vassdrags og Elektrisitetsvesen, Direktoratet for vilt og ferskvannsfisk (in Norwegian).

Jensen K.W. (1981) On the rate of exploitation of salmon from two Norwegian rivers. *International Council for the Exploration of the Sea, Copenhagen*, **C.M.1981/M:11**.

Jessop B.M. (1975) Investigation of the salmon (*Salmo salar*) smolt migration in the Big Salmon River, New Brunswick, 1966–72. *Canadian Fisheries and Marine Service, Ottawa, Technical Report Series Mar/T-75–1*, 57 pp.

Joensen J.S & Vedel Tåning Å. (1970) Marine and freshwater fishes. In *Zoology of the Faroes* (Ed. by S. Jensen *et al.*), Volumes LXII-LXIII, 241 pp.

Johnels A.G. (1984) Masken son hotar laxen (*Gyrodactylus salaris*), a parasite threatening the Atlantic salmon. *Svenskt Fiske* **9/84**, 42–4 (In Swedish).

Jonsson B & Ruud-Hansen J. (1985) Water temperature as the primary influence on timing of seaward migration of Atlantic salmon (*Salmo salar*) smolts. *Can. J. Fish, Aquat. Sci*, **42(3)**, 593–5.

Kennedy G.J.A. (1988) Stock enhancement of Atlantic salmon (*Salmo salar* L.). In *Atlantic Salmon: Planning for the Future* (Ed. by D. Mills & D. Piggins), 345–372 (The Proceedings of the Third International Atlantic Salmon Symposium, Biarritz, France). Croom Helm, London.

Kennedy G.J.A. & Greer J.E. (1988) Predation by cormorants (*Phalacrocorax carbo* L.) on the salmonid populations of the River Bush. *Aquaculture and Fisheries Management*, **19(2)**, 159–70.

Kennedy M. (1954) *The Sea Anglers' Fishes*. Hutchinson, London.

King-Webster W.A. (1969) The Galloway Dee – a short history of a salmon river. *Salmon Net*, **V**, 38–47.

Kreiberg H. (1981) Report of the joint Greenland expedition (1980), *Atlantic Salmon Trust, Farnham* 47 pp.

Landt J. (1800) Forsog til en beskrivelse over Faerøerne, Kjøbenhavn. (In Faroese.)

Laughton R. (1989) The movements of adult salmon within the River Spey. *Scottish Fisheries Research Report* **no. 41**, 19 pp.

Lear W.H. (1980) Food of Atlantic salmon in the West Greenland–Labrador Sea area. *Rapp. P.-v. Réun. Cons. int. Explor. Mer*, **176**, 55–9.

Lear W.H. & Sandeman E.J. (1980) Use of scale characters and discriminant functions for identifying continental origin of Atlantic salmon. *Rapp. P.-v. Réun. Cons. int. Explor. Mer*, **176**, 68–75.

McConnochie A.I. (1900) *Deeside*. Lewis Smith & Son, Aberdeen.

McEwan L.J. (1985) *River channel planform changes in upland Scotland with specific reference to climatic fluctuation and land-use changes over the last 250 years*. PhD thesis, University of St. Andrews.

Mackay Consultants (1989) Economic importance of salmon fishing and netting in Scotland. *A report for the Scottish Tourist Board and the Highlands and Islands Development Board, Mackay Consultants, Balloan House, Inverness*, 129 pp.

Malloch P.D.H. (1910) *Life History of the Salmon, Trout and Other Freshwater Fish*. A. and C. Black, London.

Marshall T.L. (1984) Status of Saint John River, N.B., Atlantic salmon in 1984 and forecast of returns in 1985. *CAFSAC Research Document* 84/84.

Martin J.H.A., Dooley H.D. & Shearer W.M. (1984) Ideas on the origin and biological sequences of the 1970s salinity anomaly. *International Council for the Exploration of the Sea, Copenhagen,* **C.M.1984/Gen:18**.

Martin J.H.A. & Mitchell K.A. (1985) Influence of sea temperature upon the numbers of grilse and multi-sea winter Atlantic salmon (*Salmo salar*) caught in the vicinity of the River Dee (Aberdeenshire). *Can. J. Fish. Aquat. Sci,* **42**, 1513—21.

May A.W. (1973) Distribution and migrations of salmon in the northwest Atlantic. In *International Atlantic Salmon Symposium, St. Andrews* (Ed. by M.V. Smith and W.M. Carter) **4**, pp. 373—83. International Atlantic Salmon Foundation Special Publication Series.

Meister A.L. (1962) Atlantic salmon production in Cove Brook, Maine. *Trans. Amer. Fish. Soc.,* **91**, 208—12.

Menzies W.J.M. (1914) Further notes on the percentage of previously spawned salmon. *Fisheries, Scotland, Salmon Fish., 1914,* **II**, 12 pp.

Menzies W.J.M. (1937) The movements of salmon marked in the sea. I. The North-west coast of Scotland in 1936. *Fisheries, Scotland, Salmon Fish., 1937,* **No. I**, 17 pp.

Menzies W.J.M. (1938a) Some preliminary observations on the migration of salmon on the coasts of Scotland. *Rapp. Cons. Explor. Mer.,* **108**, 18—35.

Menzies W.J.M. (1938b) The movements of salmon marked in the sea. II. West Coast of Sutherland in 1937. *Fisheries, Scotland, Salmon Fish., 1938,* **No. I**. 9 pp.

Menzies W.J.M. (1938c) The movements of salmon marked in the sea. III. The island of Soay and Ardnamurchan in 1938. *Fisheries, Scotland, Salmon Fish., 1938,* **No. VII**, 9 pp.

Menzies W.J.M. (1949) The Stock of Salmon — Its Migrations, Preservation and Improvement. *The Buckland Lectures for 1947.* Edward Arnold & Co., London.

Menzies W.J.M. & Shearer W.M. (1957) Long-distance migration of salmon. *Nature,* **179**, 790.

Millichamp R.I. (1987) *Anglers' Law.* A. & C. Black, London.

Mills C.P.R., Mahon O.A.T. & Piggins D.J. (1986) Influence of stock levels, fishing effort and environmental factors on anglers' catches of Atlantic salmon, *Salmon salar*, L., and sea trout, *Salmo trutta* L. *Aquaculture and Fisheries Management,* **17(4)**, 289—97.

Mills D.H. (1964) The ecology of the young stages of the Atlantic salmon in the River Bran, Ross-shire. *Freshwat. Salm. Fish. Res., Scotland,* **32**, 58 pp.

Mills D.H. (1986) The biology of Atlantic salmon. In *The Status of the Atlantic Salmon in Scotland* (Ed. by D. Jenkins & W.M. Shearer), 10—19 (ITE Symposium No. 15). Abbots Ripton: Institute of Terrestrial Ecology.

Mills D.H. (1987) Atlantic salmon management. In *Developments in Fisheries Research in Scotland* (Ed. by R.S. Bailey and B.B. Parrish), pp. 207—19, Fishing News Books Ltd., Farnham.

Mills D.H. (1989) *Ecology and Management of Atlantic Salmon.* Chapman and Hall Ltd., London.

Mills D.H., Griffiths D. & Parfitt A. (1978) A survey of the fresh water fish fauna of the Tweed Basin. *Nature Conservancy Council, Edinburgh,* 100 pp.

Mills D.H. & Smart N. (1982) *Report on a Visit to the Faroes.* Atlantic Salmon Trust, Pitlochry, 52 pp.

Møller Jensen J. (1980) Recaptures from the International Tagging Experiment at West Greenland. *Rapp. P.-v. Réun. Cons. int. Explor. Mer,* **176**, 122—35.

Møller Jensen J. (1988) Exploitation and migration of salmon on the high seas, in relation to Greenland. In *Atlantic Salmon: Planning for the Future* (Ed. by D. Mills & D. Piggins, 438—57, (The Proceedings of the Third International Atlantic Salmon Symposium, Biarritz, France). Croom Helm, London.

Moore A., Freake S.M. & Thomas I.M. (1990) Magnetic particles in the lateral line of the Atlantic salmon (*Salmo salar* L.). *Phil. Trans. R. Soc. Lond. B. (1990)* **329**, 11—15.

Morgan R.I.G., Greenstreet S.P.R. & Thorpe J.E. (1986) First observations on distribution, food and fish predators of post-smolt Atlantic salmon, *Salmo salar*, in the outer Firth of Clyde. *International Council for the Exploration of the Sea, Copenhagen*, **C.M./M:27**.

Munro A.L.S., Liversidge J. & Elson K.G.R. (1976) The distribution and prevalence of infectious pancreatic necrosis virus in wild fish in Loch Awe. *Proc. Roy. Soc. Edin.* **B75**, 223–32.

Munro A.L.S. & Waddell I.F. (1987) Growth of salmon and trout farming in Scotland. In *Developments in Fisheries Research in Scotland* (Ed. by R.S. Bailey and B.B. Parrish), pp. 246–63, Fishing News Books Ltd., Farnham.

Ommanney F.D. (1963) *The Fishes*. Time-Life, New York.

Österdahl L. (1969) The smolt run of a small Swedish river. In *Symposium on Salmon and Trout in Streams* (Ed. by T.G. Northcote), *H.R. MacMillan Lectures in Fisheries, University of British Columbia, Vancouver,* 205–21.

Parrish B.B. & Shearer W.M. (1977) Effects of seals on fisheries. *International Council for the Exploration of the Sea, Copenhagen*, **C.M.1977/M14**.

Payne R.H. (1980) The use of serum transferrin polymorphism to determine the stock composition of Atlantic salmon in the West Greenland fishery. *Rapp. P.-v. Réun. Cons. int. Explor. Mer*, **176**, 60–4.

Pella J.J. & Robertson T.L. (1979) Assessment of composition of stock mixtures. *Fish. Bull. U.S.* **77**, 387–98.

Piggins D.J. (1959) Investigations on predators of salmon smolts and parr. *Salmon Research Trust of Ireland. Inc.* Report and Statement of Accounts for year ended 31st December, 1958. Appendix no. 1, 12 pp.

Pippy J.H.C. (1980) The value of parasites as biological tags in Atlantic salmon at West Greenland. *Rapp. P.-v. Réun. Cons. int. Explor. Mer*, **176**, 76–81.

Pope J.A., Mills D.H. & Shearer W.M. (1961) The fecundity of Atlantic salmon (*Salmo salar* L.). *Freshwat. Salm. Fish. Res., Scotland*, **26**, 12 pp.

Potter E.C.E. & Swain A. (1982) Effects of the English north-east coast salmon fisheries on Scottish salmon catches. *Fisheries Research Technical Report*, No. 67, Ministry of Agriculture, Fisheries and Food, Lowestoft, 8 pp.

Pratten D.J. & Shearer W.M. (1981) The fishing mortality of North Esk salmon. *International Council for the Exploration of the Sea, Copenhagen*, **C.M.1981/M:26**.

Pyefinch K.A. (1952) Capture of the pre-grilse stage of salmon. *Scott. Nat.*, **64**, 47.

Pyefinch K.A. & Shearer W.M. (1957) The movements of salmon tagged in the sea, Altens, Kincardineshire 1952. *Freshwat. Salm. Fish. Res., Scotland*, **19**, 7 pp.

Pyefinch K.A. & Woodward W.B. (1955) The movements of salmon tagged in the sea, Montrose, 1948, 1950, 1951. *Freshwat. Salm. Fish. Res., Scotland*, **8**, 15 pp.

Radford A. (1984) The economics and value of recreational salmon fisheries in England and Wales: An analysis of the rivers Wye, Mawddach, Tamar and Lune. *Marine Resources Research Unit, Portsmouth Polytechnic*, 250 pp.

Rae B.B. (1960) Seals and Scottish fisheries. *Marine Research Series, Scotland*, **2**, 39 pp.

Rae B.B. (1964) News items. *Scott. Fish. Bull.* **22**, 17.

Rae B.B. (1966) News items. *Scott. Fish. Bull.* **25**, 33.

Rae B.B. (1969) The food of cormorants and shags in Scottish estuaries and coastal waters. *Marine Research Series, Scotland*, **1**, 16 pp.

Rae B.B. & Shearer W.M. (1965) Seal damage to salmon fisheries. *Marine Research Series, Scotland*, **2**, 39 pp.

Randall R.G. (1984) Number of salmon required for spawning in the Restigouche River, N.B., *CAFSAC Research Document* 84/16, 15 pp.

Reddin D.G. (1985) Contribution of North American salmon to the Faroes fishery.

International Council for the Exploration of the Sea, Copenhagen, **C.M.1985/M:11**.

Reddin D.G. (1988) Ocean life of Atlantic salmon in the Northwest Atlantic. In *Atlantic Salmon: Planning for the Future* (Ed. by D. Mills and D. Piggins), pp. 483–511 (The Proceedings of the Third International Atlantic Salmon Symposium, Biarritz, France). Croom Helm, London.

Reddin D.G. & Carscadden J.E. (1982) Salmon-capelin interactions. *International Council for the Exploration of the Sea, Copenhagen*, **C.M.1982/M:17**.

Reddin D.G. & Shearer W.M. (1987) Sea-surface temperature and distribution of Atlantic salmon (*Salmo salar L.*) in the northwest Atlantic. *American Fisheries Society Symposium* 1, 262–75.

Rice D.E.A. (1988) The landscape of the Spey catchment. In *Land Use in the River Spey Catchment* (Ed. by D. Jenkins), pp. 155–161 (ACLU Symposium No.1) Aberdeen: Aberdeen Centre for Land Use.

Ritter J.A. (1975) Lower ocean survival rates for hatchery reared Atlantic salmon (*Salmo salar L.*) stocks released in rivers other than their native streams. *J. Cons. Perm. Int. Explor. Mer.*, **26**, 1–10.

Rommel S.A. Jr. & McLeave J.D. (1973) Sensitivity of American eels (*Anguilla rostrata*) and Atlantic salmon (*Salmo salar*) to weak electric and magnetic fields. *J. Fish. Res. Bd Can.*, **30**, 657–63.

Rosenørn S., Fabricius J.S., Buch E. & Horsted Sv. Aa. (1985) Record hard winters at West Greenland. *North Atlantic Fisheries Organization, Secretariat. Doc.* 85/61, Ser. No. N1011, 18 pp.

Ross R.M. (1986) The law relating to salmon fishing. In *The Status of Atlantic Salmon in Scotland* (Ed. by D. Jenkins and W.M. Shearer), pp. 3–9, (ITE Symposium No.15). Abbots Ripton: Institute of Terrestrial Ecology.

Rosseland L. (1979) Litt om bestand og bestatning av laksen fra Laerdalselva. In *Vassdragsregulingers biologiske virkninger i magasiner og lakseelver* (Ed. by T.B. Gunnerod & P. Mellquist), pp. 174–86, Norges Vassdrags og Elektrisitetsvesen, Direktoratet for vilt og ferskvannsfisk. (In Norwegian.)

Royce W., Smith L.S. & Harth A.C. (1968) Models of oceananic migrations of Pacific salmon and comments on guidance mechanisms. *Fish. Bull.* **66**, 441–62.

Russel A. (1864) *The Salmon*. Edmonston and Douglas, Edinburgh.

Russell I.C. & Buckley A. (1989) Salmonid and freshwater fisheries statistics for England and Wales, 1987. *Fish. Res. Data Rep., MAFF Direct. Fish. Res., Lowestoft*, (16), 27 pp.

Saila S.B. & Shappy R.A. (1963) Random movement and orientation in salmon migration. *J. Cons. Int. Explor. Mer 28*, 153–166.

Scarnecchia D.L. 1984. Climatic and oceanic variations affecting yield of Icelandic stocks of Atlantic salmon (*Salmo salar*). *Can. J. Fish. Aquat. Sci.*, **41**, 917–35.

Schaefer M.B. (1951) Estimates of the size of animal populations by marking experiments. *Fishery Bull., Fish Wildl. Serv. U.S.* **52**, 189–203.

Sedgwick S.D. (1970) Report of the Inspector of Salmon and Freshwater Fisheries. *Annual Report of Fishery Board for Scotland, 1969*, 41–56.

Shearer W.M. (1958) The movement of salmon tagged in the sea, Montrose, 1954, 1955. *Freshwat. Salm. Fish. Res.*, **20**, Scotland, 11 pp.

Shearer W.M. (1972) *A study of the Atlantic salmon population in the North Esk, 1961–70*. M.Sc. thesis, University of Edinburgh, 437 pp.

Shearer W.M. (1984a) The relationship between both river and sea-age and time of return to homewaters in Atlantic salmon. *International Council for the Exploration of the Sea. Copenhagen*, **C.M.1984/M:24**.

Shearer W.M. (1984b) The natural mortality at sea for North Esk salmon. *International Council for the Exploration of the Sea. Copenhagen*, **C.M.1984/M:23**.

Shearer W.M. (1985a) Scottish salmon catches 1952–81. *Salmon Net*, **XVIII**, 66–71.

Shearer W.M. (1985b) Salmon catch statistics for the River Dee, 1952−83. In *The Biology and Management of the River Dee* (Ed. by D. Jenkins), pp. 127−141 (ITE Symposium No. 14). Abbots Ripton: Institute of Terrestrial Ecology.

Shearer W.M. (1986a) The exploitation of Atlantic salmon in Scottish homewater fisheries in 1952−83. In *The Status of the Atlantic Salmon in Scotland* (Ed. by D. Jenkins and W.M. Shearer), pp. 37−49 (ITE Symposium No. 15) Abbots Ripton: Institute of Terrestrial Ecology.

Shearer W.M. (1986b) An evaluation of the data available to assess Scottish salmon stocks. In *The Status of the Atlantic Salmon in Scotland* (Ed. by D. Jenkins and W.M. Shearer), pp. 91−111 (ITE Symposium No. 15). Abbots Ripton: Institute of Terrestrial Ecology.

Shearer W.M. (1988a) Fisheries in the Spey catchment. In *Land Use in the River Spey catchment* (Ed. by D. Jenkins), 197−212. (ACLU Symposium No. 1.) Aberdeen: Aberdeen Centre for Land Use.

Shearer W.M. (1988b) Long term fluctuations in the timing and abundance of salmon catches in Scotland. *International Council for the Exploration of the Sea, Copenhagen.* **C.M.1988/M:21**.

Shearer W.M. (1988c) Spey Research Project. Summary of Results 1983−1985. In *Spey District Fishery Board Report 1984−87, Schedule 2*, 3 pp.

Shearer W.M. (1989) The River Tweed Salmon and Sea Trout Fisheries. In *Tweed Towards 2000* (Ed. by D. Mills, 60−79) (Tweed Foundation Symposium No. 1) Tweedmouth, The Tweed Foundation.

Shearer W.M. (1990) The Atlantic salmon (*Salmo salar* L.) of the North esk with particular reference to the relationship between both river and sea age and time of return to home waters. *Fisheries Research*, **10 (1990)**, 93−123.

Shearer W.M. & Balmain K.H. (1967) Greenland Salmon. *Salmon Net*, **III**, 19−24.

Shearer W.M., Cook R.M., Dunkley D.A., MacLean J.C. & Shelton R.G.J. (1987) A model to assess the effect of predation by sawbill ducks on the salmon stock of the River North Esk. *Scottish Fisheries Research Report No. 37*, 12 pp.

Shelton R.G.J. & Dunkley D.A. Stock assessment and the Atlantic salmon, *Salmo salar* L. (in press).

Smart G.G.J. (1965) The practical results of increasing the accessibility of spawning grounds in the North Esk. *The Salmon Net*, **I**, 21−25.

Smith G.W. (1990) The relationship between river flow and net catches of salmon (*Salmon salar* L.) in and around the mouth of the Aberdeenshire Dee between 1973 and 1986. *Fisheries Research*, **10 (1990)**, 73−91.

Smith G.W., Hawkins A.D., Urquhart G.G. & Shearer W.M. (1981) Orientation and energetic efficiency in the offshore movements of returning Atlantic salmon, *Salmo salar* L. *Scottish Fisheries Research Report No. 21*, 22 pp.

Solomon D.J. (1984) *Salmonid enhancement in North America. Description of some current developments and their application to the U.K.* Atlantic Salmon Trust, Pitlochry, 40 pp.

Solomon D.J. (1985) Salmon stock and recruitment. *J. Fish. Biol.*, **27**, **Suppl. A**, 45−57.

Stansfeld J.R.W. (1989) Is a rod-caught salmon more valuable to the Scottish economy than one caught by net? *The Salmon Net*, **XXI**, 59−64.

Stasko A.B., Sutterlin A.M., Rommel S.A. & Elson P.F. (1973) Migration-orientation of Atlantic salmon (*Salmo salar* L.) In *International Atlantic Salmon Symposium, St Andrew's.* (Ed. by M.W. Smith & W.M. Carter), *International Atlantic Salmon Foundation Special Publication Series*, *4(1)*, 119−37.

Stewart L. (1978) Why no southern hemisphere salmon? The gyre theory. *Salmon and Trout magazine*, **No. 213**, 46−50.

Struthers G. (1970) A report on a salmon long-lining cruise off the Faroes during April 1970. *Freshwater Fisheries Laboratory, Pitlochry*, Report **54FW70**.

Struthers G. (1971) A report on the 1971 long-lining cruise off the Faroes. *Freshwater Fisheries Laboratory, Pitlochry*, Report **33FW71**.

Struthers G. (1981) Observations on Atlantic salmon (*Salmo salar* L.) stocks in the sea off the Faroe Islands 1969−79. *Working document to the North Atlantic Salmon Working Group*, 11 pp.

Struthers G. & Stewart D. (1986) Observations on the timing of migration of smolts from natural and introduced juvenile salmon on the upper River Tummel, Scotland. *International Council for the Exploration of the Sea, Copenhagen*, **C.M.1986/M:4**.

Svabo J. Chr. (1782). Indberetninger fra en Reise i Foeroe 1781 og 1782. *N. Djurhuus, København 1959*. (In Danish).

Symons P.E.K. (1979) Estimated escapement of Atlantic salmon (*Salmo salar*) for maximum smolt production in rivers of different productivity. *J. Fish. Res. Bd Can.*, **36**, 132−40.

Templeman W. (1967) Atlantic salmon from the Labrador Sea and off West Greenland taken during 'A.T. Cameron' Cruise, July−August 1965. *Bull. int. Comm. NW Atlant. Fish.*, **No. 4**, 5−40.

Thomas T.B. (1964) Fishing from early times. *The Atlantic Salmon Journal*, **13**, No. 1, 5−6.

Thomson J.M. (1979) The Stormontfield piscicultural experiments 1853−1866. *Salmon Net*, **XII**, 31−33.

Thorpe J.E. (1988) Salmon enhancement: Stock discreteness and choice of material for stocking. In *Atlantic salmon: Planning for the Future* (Ed. by D. Mills and D. Piggins), pp. 373−83. (The Proceedings of the Third International Atlantic Salmon Symposium, Biarritz, France). Croom Helm, London.

Trench C.C. (1974) *A history of angling.* Follett Publishing Company, Chicago.

Tytler P., Thorpe J.E. & Shearer W.M. (1978) Ultrasonic tracking of the movements of Atlantic salmon smolts (*Salmo salar* L.) in the estuaries of two Scottish rivers. *J. Fish Biol.*, **12**, 575−86.

Veitch A. (1989) Illegal fishing. In *Tweed Towards 2000* (Ed. by D. Mills, 80−4) (Tweed Foundation Symposium No.1). Tweedmouth, The Tweed Foundation.

Vibert R. (1952) Migrations maritimes de saumons. *Bull. Soc. Zool. Fr.*, **77**, 313−7.

Walker J. (1988) A wake for the salmon. *Northumberland County Library, Berwick-on-Tweed*, 47 pp.

Walton Izaak (1653) *The Compleat Angler*, Marriott, London.

Webb J. (1989) The movements of adult salmon in the River Tay. *Scottish Fisheries Research Report*, **No. 44**, 32 pp.

Webb J. & Hawkins A.D. (1989) The movements and spawning behaviour of adult salmon in the Girnock Burn, a tributary of the Aberdeenshire Dee, 1986. *Scottish Fisheries Research Report*, **No. 40**, 41 pp.

Westerberg H. (1982) Ultrasonic tracking of Atlantic Salmon. I. Movements in coastal regions. II. Swimming depth and temperature stratification. *Institute of Freshwater Research, Drottningholm, Report*, **60**, 81−120.

Westerberg H. (1984) The orientation of fish and vertical stratification at fine- and micro-structure scales. In *Mechanisms of Migration in Fishes* (Ed. by J.D. McCleave, G.P. Arnold, J.J. Dodson and W.H. Neill, 179−204), Plenum Publishing Corporation, New York.

Wheeler A. & Gardner D. (1974) Survey of the literature of marine predators on salmon in the north-east Atlantic. *Fish. Mgmt.*, **5(3)**, 63−6.

Whelan B.J. & Marsh G. (1988) An economic evaluation of Irish angling. *A report prepared for the Central Fisheries Board by the Economic and Social Research Institute, Dublin.*

Younger J. (1840) *On River Angling for Salmon and Trout.* Rutherford, Kelso.

The model used to estimate the natural mortality at sea for North Esk salmon

The first step in the analysis involves estimating the number of salmon ascending the R. North Esk during the fishing season (N) from:

$$N = N_d - (C_d + C_g) + C_r \tag{1}$$

where C_r = North Esk net and coble catch, C_d = net and coble catch at Morphie Dyke (the furthest upstream major net fishery in the R. North Esk), C_g = net and coble catch at the Gauge (a small net fishery and the only one upstream of Morphie Dyke), and N_d = estimated population reaching Morphie Dyke during the netting season.

In this analysis, compared with Pratten & Shearer (1981) and Shearer (1984b), N_d has been recalculated using the method proposed by Schaefer (1951) for estimating stratified populations.

The numbers of North Esk fish caught in net fisheries in Scotland but outside the R. North Esk (C_c) were estimated by:

$$C_c = \frac{M_c \cdot C_r}{M_r} \tag{2}$$

where M_r = number of tagged fish in North Esk net and coble catch, and M_c = number of tagged fish in Scottish net fisheries outside the R. North Esk.

The total number of fish returning to Scottish home-waters from any smolt year (R_s), assuming no losses due to natural mortality in home-waters or to illegal fishing, was estimated by:

$$R_s = \frac{N}{(1-E)} + C_c \tag{3}$$

It has been assumed that tag loss after the fish are available to fisheries is negligible. This seems to be a reasonable assumption because the examination of fish which have returned lacking tags showed that tag loss occurs soon after the smolts have entered the sea. This is proved both by the absence of any wound consistent with a tag recently having been lost, and by regenerated scales from the area where the tag was initially inserted showing the pattern associated with the sea life of the fish.

Because there were no data on the number of fish returning outside the fishing season, Pratten & Shearer (1981) and Shearer (1984b) repeated the calculations using E = 0.10 and E = 0.20 in order to estimate the numbers of salmon returning to home-waters. Recent investigations on the R. North Esk using an electronic fish counter showed that more realistic values for 1SW and MSW salmon are E = 0.47 and 0.10 respectively. It is likely that there will be additional losses after salmon return to Scottish home-waters as a result of natural mortality (M) and illegal fishing (C_i). For these reasons the total number returning from any smolt year (R_t) is given by:

$$R_t = R_s + M + C_i \tag{4}$$

However, no data presently exist to allow either M or C_i to be quantified.

The model used to assess losses to home water stocks resulting from the Faroese fishery

Because the losses to home water stocks are estimated separately for each smolt year class returning to home waters after spending different periods in the sea, the equation used to estimate total short-term losses (LF) is:

$$LF = \frac{1}{(1-N)}\, \Sigma_{ij}\, (WR_{ij} \times PW_i \times PN_{ij} \times S_{ij})$$

where subscript 'i' refers to sea age classes (1 October is the nominal birthday) of salmon taken in the Faroes fishery (discards are treated as a separate sea age class) and subscript 'j' refers to the sea age classes of the same stocks on their return to home waters. The parameters are then defined as follows:

N = non-catch fishing mortality expressed as a proportion of the total fishery-induced mortality

WR_{ij} = ratio of the weight of each sea age class in home waters to their mean weight in catches at the Faroes

PW_i = proportion by weight of each sea age class relative to the total nominal catch.

PN_{ij} = estimated proportions of the fish of each sea age class in the fishery returning in the same and subsequent years

S_{ij} = survival rates of different sea age classes between the Faroes fishery and home waters

Errors

The 1984 report of the ICES Working Group on North Atlantic Salmon (Anon 1984a) gives a useful summary of the likelihood of error in the separate input parameters for a major assessment (the 1984 assessment of the Faroese fishery) and the relative effect of errors on results. The main possible errors include the following:

Discards

Discarded fish contribute only 2% of the total loss estimate, and the discard rate (5.5%) is thought to be well estimated. Thus, errors in this parameter are expected to have virtually no effect on the overall assessment. Any change in the discard rate is likely to be related to a change in the age composition of the catch, which would itself have a greater effect on the result.

Non-catch fishing mortality (N)

The parameter (a nominal value of 0.1 was assumed by the Working Group) has a proportional effect on the total assessment. If N is assigned a value of 0.15, the estimate of total loss is increased by 6% of the estimated value; and if N is assigned a value of 0.05 the estimate of total loss is decreased by 5%.

Proportion by weight in each sea age class (PW$_i$)

The estimates of PW$_i$ for the 1982−3 season are based on extensive sampling and are thought to be trustworthy. Changes may occur in the age composition of the population in the fishery area in different years. However, even a substantial change from a catch comprising 6%, 78% and 16% of 1SW, 2SW and 3SW fish to one of 30%, 65% and 5% respectively gives only a 5% increase in the assessment result.

Proportion returning in the same year (PN$_{ij}$)

It was suggested that the serum steroid sampling method used to estimate this parameter (at 78%) was more likely to underestimate than overestimate the true value. Using the value of 90% derived from the tagging experiments in the early 1970s reduces the estimated loss by 6%.

Weight ratios (WR$_{ij}$)

There was considerable variation (up to ±15%) in the values used to give the average weight ratios for each sea age class. In the unlikely event that all values were biased in one direction, the final assessment would be increased or decreased by the same proportion. Errors in the weight parameters for the 2SW fish would have a disproportionate effect on the assessment.

Survival (S$_{ij}$)

The model previously used to estimate monthly M is based on very limited data and may be inaccurate. A value of 0.01 was assumed for all sea age classes assessed. If M is increased to 0.015, the estimated loss would be decreased by 3%. If M is decreased to 0.005, the estimated loss would be increased by 3%. Alternatively, a 5% additional natural mortality on homing fish, caused by straying or predation in home waters, would decrease the assessment by 5%.

APPENDIX C

Estimating the effect of the removal of net fisheries

Here, I have chosen an imaginary river somewhere in Scotland, but not on the west coast, with fixed engine fisheries within the fishery district and in adjacent districts, and rod and line and net and coble fisheries within estuarial limits. The mean fixed engine, net and coble and rod and line grilse and salmon catches for the last five years were

net and coble	5500	3500−8400
rod and line	1400	870−1800

It is worth noting the wide range of catches by all three gears.

If all netting ceases within this fishery district, the possible effects can be considered under three broad headings:

- Rod catch
- Spawning stock
- Smolt production

Rod catch

It cannot be assumed in any fishery district, that the total net catch (fixed engine and net and coble) will automatically become available to the rod fishery in the absence of netting for the following reasons:

(1) The fixed engine catch

    (a) A proportion (about 35%) of this catch will not be native to the river.
    (b) It is reasonable to assume that losses to predators and to poachers will increase by about 5% in the absence of active coastal net fisheries.
    (c) Losses (about 5%) will occur to coastal net fisheries still operating in other fishery districts. (Some of the fish which would have been caught if the net fishery had not been removed will migrate outwith the fishery district containing this imaginary river and a proportion will be caught in the coastal net fisheries in other fishery districts.)

(2) The net and coble catch

    (a) A proportion (about 24%) of this catch will not be native to the river.

(b) It is reasonable to assume that losses to predators and to poachers will increase by about 5% in the absence of an active net fishery.

These assumptions suggest that from a combined fixed engine and net and coble catch of 11 000 salmon and grilse, an additional 7315 fish might become available to the rod fishery. Provided that these additional fish do not alter the general catchability of fish in the river, the average rod catch could be expected to increase by some 732 fish (based on an exploitation rate of 10%) from 1400 to 2132. Because most of the additional fish are likely to be grilse (during the previous five years, 80% of the commercial catch were identified as grilse), the availability of these additional fish to the rod fisheries sited along the river would decrease with distance upstream from the river mouth. The beats nearest the mouth would benefit most and those nearer the head-waters least, if at all (Shearer 1984a and 1988b). In addition, it is recognized by many salmon authorities that grilse are notoriously more difficult to catch on rod and line than salmon (Sedgwick 1970).

The spawning stock

The numbers of additional potential spawners would be:

7315 − 732 = 6583 fish.

The fish escaping capture by the rod fisheries less losses to poachers and natural causes would be additional potential spawners.

Based on the most recent data from the R. North Esk, a 10% loss to poaching and natural causes would not be unreasonable, although losses between years vary widely. Nevertheless, the very presence of additional fish in the river as a direct result of the cessation of netting, could increase natural mortality above the value of 10%, particularly if drought conditions prevailed, and the salmon were congregated below obstructions for an extended period. Thus a 90% survival is probably nearer the maximum than the minimum value. Assuming a value of 10% for mortality, the number of additional spawners would be:

6583 − 658 = 5925 fish.

Based on the results from sampling and ageing the commercial catch prior to the removal of the net fishery, the data shown in Table C1 were obtained.

Table C1

	Five-year mean	
	MSW fish	1SW fish
Percentage sea age composition	20	80
Mean length (cm)	78	61
Sex ratio (M:F)	1.00:1.75	1.00:0.74

Table C2

	MSW	1SW
Male	431	2724
Female	754	2016
Total	1185	4740

Table C2 gives the estimated number of additional MSW and 1SW fish, broken down into males and females, in the spawning stock which would result from the cessation of netting.

Smolt production

Buck & Hay (1984) examined the relationship between the estimated egg production and the number of smolts produced in the Girnock Burn and suggested a survival rate of 1% between egg and smolts. A similar study on the R. North Esk produced a mean survival value of 0.8%, which is not significantly different from the value obtained on the Girnock Burn and the same as that published for the Pollet River in Canada (Symons 1979). The additional egg deposition from the 754 female MSW and 2016 female 1SW fish would be 10 955 754 of which 1% (109 557) might survive to the smolt stage. However, this assumes that the egg deposition prior to the removal of the net fishery was limiting smolt production.

In all these studies, it is assumed that the survival rate of juvenile salmon is density-dependent and that egg depositions above the optimum number for a particular spawning burn or river would not result in an increased smolt production (Beverton & Holt 1957). On the other hand, if egg deposition previously limited smolt production, any increase in egg deposition could be expected to produce more smolts. Buck & Hay (1984) suggested that an egg deposition of 2.6 m^{-2} was necessary to reach optimum production in the Girnock Burn and that egg depositions in excess of 3.4 m^{-2} would not result in an increased migrant parr population. Comparable Canadian data (Symons 1979) suggested somewhat similar optimum egg deposition levels of 1.65 m^{-2} − 2.20 m^{-2}.

If optimum egg deposition can be obtained despite a net fishery, an increase in egg deposition following the removal of that fishery may not result in more smolts.

APPENDIX D

The values or reliable estimates which are required to construct a management plan and monitor the results

The values/estimates are as follows:

(1) Number, age (river and sea) composition, length frequency distribution and sex of ascending adults. (A reliable fish counter is now available and a trap for sampling migrants can be incorporated into the weir structure which supports the electrode array.)
(2) The number, age-composition, length frequency distribution and sex ratio of the catch taken at each fishery.
(3) Losses after the fish are counted and before they spawn. The length of each fish, its sex and a sample of scales.
(4) The length fecundity relationship of each stock component.
(5) The annual smolt production together with the age composition of the emigrants. (The trap incorporated in the weir can be used to provide the necessary material.)
(6) A measure of the juvenile rearing habitat to allow the number of spawners necessary to seed it at the preferred egg density to be estimated.

Index

Aberdeen, 59, 131, 133
Aberdeen Harbour Board, 204
acid rain, 197
acidification, 188, 194, 197–8
Acts of Parliament
  Freshwater and Salmon Fisheries
    (Scotland) Act 1976, 85
  Leases Act of 1449, 85
  River Boards Act 1948, 93
  Salmon Act 1986, 90–91, 94–5, 151, 197
  Salmon and Freshwater Fisheries
    Act 1907, 93
  Salmon and Freshwater Fisheries
    Act 1923, 93, 172
  Salmon and Freshwater Fisheries
    Act 1972, 173
  Salmon and Freshwater Fisheries
    Act 1975, 88, 93, 106
  Salmon and Freshwater Fisheries
    (Protection) (Scotland)   Act 1951,
    90, 151–2
  Salmon Fisheries Act 1861, 92
  Salmon Fisheries Act 1865, 92
  Salmon Fisheries Act 1873, 92
  Salmon Fisheries (Scotland) Act 1828, 89
  Salmon Fisheries (Scotland) Act 1844, 89
  Salmon Fisheries (Scotland) Act 1862,
    89–90
  Salmon Fisheries (Scotland) Act 1863, 90
  Salmon Fisheries (Scotland) Act 1864, 90
  Salmon Fisheries (Scotland) Act 1868, 90
  Sea Fish Industry Act 1959, 99
  Sea Fish Industry Act 1962, 99
  Tweed Fisheries Act 1857, 91
  Tweed Fisheries Act 1859, 91
  Tweed Fisheries Act 1969, 91
  Water Act 1989, 94
Advantage gained by changing from bag
  and stake

  nets to fixed gill nets, 202–203
age/weight key, 170
alder *Alnus glutinosa*, 10
alevin, 5, 207
aluminium, 194, 197
American merganser, *Mergus merganser
  americanus*, 52
anadromous, 2, 5
angling effort, 159–60
angling methods, 110
annual close season *see* annual close time
annual close time, 22, 25, 79–80, 90–95,
  111–12, 138, 145, 211
annual close time Order, 90, 201
Aquantic Ltd, 17
Ardnamurchan, 131
Arran, 131
Association of Scottish District Salmon
  Fishery Boards, 88
Atlantic Ocean, 40, 73, 210
automatic fish counter, 8, 16–19, 41, 79,
  141, 145, 148, 156, 201–202
availability to nets, 135, 143
Avon and Dorset Fishery Board, 177

Bag nets, 12, 21, 72, 91, 97–100, 102,
  105–106, 115, 118, 131, 136, 139, 151–2,
  183, 202–203
Ballater, 95
Ballindalloch, 110
Baltic Sea, 2, 21, 189
basal hormone levels, 62
basket fishery, 98
basket weirs, 108
Battle of Tenerife, 109
Bay of Biscay, 2
Bay of Fundy, 59
beech *Fagus sylvatica*, 10
Bellenden, John, 20

'Baulk', 108
effort, 158–60
'Garth', 108
mortality, 79, 148
rights
  England and Wales, 88
  Scotland, 85–7
rod, 109
season, 22, 91–2, 101, 110, 141, 155,
  157–8, 170–71
tackle, 109, 157
weir, 108
fixed engines *see* bag nets; stake nets
fixed nets, 58, 97, 105, 205
floating long lines *see* pelagic long lines
fly nets *see* stake nets
Fochabers, 54
Forestry Commission, 193–4
France, 66
freshets, 21
Freshwater Fisheries Laboratory, 203
fry
  densities, 42
  dispersal, 43
  feeding behaviour, 46–7
  mortality, 45
  predation, 45
  starvation, 45
  survival, 45
Fuglø Head, 70
furunculosis, 189, 191

gaffs, 98
Gairloch, 188, 191
Garmouth, 85
gene bank, 188, 191, 198
gene pool, 188, 190
geographic regions, 152
gill net, 102, 106, 120–21, 174, 203
glass reinforced plastic (GRP), 17
Golden Sea Produce, 186
gossander *Mergus merganser*, 52
Grantown, 110
'Greenhouse' effect, 198–9
Greenland, 2, 5, 64–5, 70, 73, 83
Greenland (east)
  countries contributing to fishery, 65
  total catch by weight, 122
Greenland (west)

assessment of effect of fishery on home
  water stocks, 125–6
assessment of loss to angling catches,
  126–7
catch quota (TAC), 122
countries contributing to salmon catch,
  66
exploitation rate, 122
fishing
  grounds, 120
  methods, 120–21
  regulations, 122
  season, 120–21
low catches, 67–70, 200
mean weight of salmon caught, 67
proportion of North American and
  European fish in catches, 67–8, 122
recapture sites of tagged fish, 59
salmon catches, 121
sea age of the catch, 67
sex ratio in the catch, 67
the fishery, 4, 59, 120–22, 129, 200, 209,
  212
total catch by weight, 67–9, 121

grey seal *Halichoerus grypus*, 77–9,
  205–206
'grilse' error, 155, 166
grilse salmon ratio, 23, 25, 34, 49, 70, 73, 80
Grantown, 110
Grimsey Island, Iceland, 70
Grontveld Brothers, 186
gyre, 73
Gyrodatylus salaris, 189, 191

haaf net, 101–102, 151
Halliburton, John, 110
hand nets, 108
Harris, 112
heave net *see* haaf net
Helmsdale, 133
Henderson, W., 164
herring *Clupea harlengus*, 58, 71–2
high seas fisheries *see* Faroes, Fishery for
  salmon in international waters, East and
  West Greenland, and Norwegian Sea
Highland Boundary Fault, 10
Highlands and Islands Development Board,
  112
Hogarth, 115